Not Too Late

Edited by Rebecca Solnit & Thelma Young Lutunatabua

Contributors:

Julian Aguon
Jade Begay
adrienne maree brown
Edward R. Carr
Renato Redentor Constantino
Joëlle Gergis
Jacquelyn Gill
Roshi Joan Halifax
Mary Annaïse Heglar
Mary Anne Hitt
Nikayla Jefferson

Kathy Jetñil-Kijiner
Antonia Juhasz
Fenton Lutunatabua
Yotam Marom
Denali Sai Nalamalapu
Joseph Zane Sikulu
David Solnit
Leah Cardamore Stokes
Farhana Sultana
and
Gloria Walton

Also Available from Haymarket Books by Rebecca Solnit

Call Them by Their True Names: American Crises (and Essays)

Cinderella Liberator, illustrated by Arthur Rackham

Hope in the Dark: Untold Histories, Wild Possibilities

Los hombres me explican cosas

Men Explain Things to Me

The Mother of All Questions

Waking Beauty, illustrated by Arthur Rackham

Whose Story Is This?: Old Conflicts, New Chapters

"City of Women" map of New York City,
created with Joshua Jelly-Schapiro

"City of Women" map of London,
created with Reni Eddo-Lodge and Emma Watson

Not Too Late

CHANGING THE CLIMATE STORY
FROM DESPAIR TO POSSIBILITY

Edited by
Rebecca Solnit &
Thelma Young Lutunatabua
with illustrations by David Solnit

Haymarket Books
Chicago, Illinois

Published in 2023 by
Haymarket Books
P.O. Box 180165
Chicago, IL 60618
773-583-7884
www.haymarketbooks.org
info@haymarketbooks.org

ISBN: 978-1-64259-897-1

Distributed to the trade in the US through Consortium Book Sales and
Distribution (www.cbsd.com) and internationally through Ingram Publisher Services International (www.ingramcontent.com).

This book was published with the generous support of Lannan Foundation
and Wallace Action Fund.

Special discounts are available for bulk purchases by organizations and institutions. Please email info@haymarketbooks.org for more information.

Cover design by Abby Weintraub.

Printed in Canada by union labor.

Library of Congress Cataloging-in-Publication data is available.

10 9 8 7 6 5 4 3 2 1

Contents

FRAMEWORKS OF POSSIBILITY

THE FUTURE WE WANT

Having hope and maintaining hope is a chore. And that's something we should be honest about. Right, it's work. It is not easy to be hopeful all the time. That's the beautiful part about having people around you who are encouraging and who are constantly reminding you that you are built for this moment, that you are meant for this moment, that you're right for this moment.

—**Tarana Burke, in conversation with Ai-jen Poo**

You are not obligated to complete the work, but neither are you free to abandon it.

—**from the Pirkei Avot**

I don't feel that I have the right to consider giving up hope . . . I appreciate Mariame Kaba's idea that hope is a discipline. It's a choice—it can't be a matter of fluctuating affect, whatever viral news story or TikTok gave you hope in people or took it away. In general, I try to expect nothing and hope that everything is possible. I want the courage to need very little and demand a lot.

—**Jia Tolentino**

Difficult Is Not the Same as Impossible

Rebecca Solnit

It is late. We are deep in an emergency. But it is not too late, because the emergency is not over. The outcome is not decided. We[1] are deciding it now. The longer we wait to act, the more limited the options, but scientists tell us there are good options and great urgency to embrace them while we can. An emergency is when a stable situation destabilizes, when the house catches fire or the dam breaks or institution implodes, when the failure or sudden change or crisis calls for urgent response. It's when it becomes clear that the way things were is not how they're going to be.

---

1. The word "we" is both problematic and necessary, so at the outset I want to acknowledge that not everyone is part of any version of "we." This book was put together with young people and newcomers to the climate movement in mind, with an expectation that most readers would be in the US and Global North. Even there, the differences matter—between Indigenous and settler, rich and poor, people who have lost homes and lives to the climate crisis and those who think of it as largely in the future. But there's also the "we" that is all humanity, all who are impacted, and all beings alive now and yet to come. So, take this "we" with a grain of salt and allow some latitude for the necessity and inadequacy of the categories that make up language.

Some days it can feel like you're the house that caught fire, but you might also turn out to be the firefighter or the water. You might even be the other interpretation that comes along to say some structures should be burned to the ground because they were prisons or miseries. An emergency can involve terrible loss or it can bring about magnificent transformation, and, while it's unfolding, the outcome can be impossible to foresee. Or it can depend on what you and we do.

The word itself comes from *emerge*, to exit, to leave behind, separate yourself from, so an emergency is when you exit from the familiar and the stable. We are now doing that on a planetary scale, exiting the stability and reliability of the delicately tuned systems of seasons, weather, relations between species, even the shape of the world for the past ten thousand years, as oceans draw new coastlines, lakes dry up, glaciers melt, and natural and human communities shift and migrate. It is an exodus into the unknown, and our task is to make a home there for ourselves and for the nature from which we were never separate.

We are deep in an emergency, and we need as many people as possible to do what they can to work toward the best-case scenarios and ward off the worst. Involvement depends on having a sense of personal power—the capacity to make an impact. Inseparable from that sense is the hope that it matters that you do it. Many in desperate circumstances have believed that it matters even when they didn't believe they could win (and sometimes they won anyway, or they inspired someone else to win). A lot of stories in circulation endeavor to strip you of hope and power, to tell you it doesn't matter or it's too late or there's nothing you can do or we never win. *Not Too Late* is a project to try to return hope and power through both facts and perspectives.

Twenty years ago, I began to speak directly about hope and to what impeded it for so many people: "I say all this to you because hope is not like a lottery ticket you can sit on the sofa and clutch, feeling lucky. I say this because hope is an ax you break down doors with in an emergency." I wrote during a surge of neoliberal attacks on indigenous people, nature, small farmers, and labor, during the so-called Global War on Terror of the oil-fueled, climate-denying Bush administration that killed more than half a million and demonized billions. People pushed back, there were victories, and they mattered.

There was virtually no climate movement to speak of, though there were campaigns against the coal and oil industry and their environmental destruction and human rights violations, and Indigenous people had never stopped striving to care for land, water, and nature. The world is now both better and worse than we imagined twenty years ago. The parts that are better are better largely because of grassroots campaigns, popular movements, and Indigenous uprisings by people who believed it was worth trying to act on their beliefs and commitments.

These are connected to the way that new ideas about justice, equality, kindness, about interdependence as the first lesson nature teaches us, appeared on distant horizons like clouds and then soaked into the soil of the collective imagination like spring rain. (Many of the new ideas were really old, discarded, discredited ideas that came back because we wanted or needed them or were more able to hear the people who had never forgotten them.) The parts that are worse—in many, perhaps most cases, they're worse due to good people doing nothing, or rather not enough of them trying.

Hope is not optimism. Optimism assumes the best, and assumes its inevitability, which leads to passivity, as do the pessimism and cynicism that assume the worst. Hope, like love, means taking risks and being vulnerable to the effects of loss. It means recognizing the uncertainty of the future and making a commitment to try to participate in shaping it. It means facing difficulties and accepting uncertainty. To hope is to recognize that you can protect some of what you love even while grieving what you cannot—and to know that we must act without knowing the outcome of those actions.

Over and over again, the world has been changed by people who, at the outset, seemed far too puny to pit themselves against the most powerful institutions of their time. After centuries of genocide, Indigenous people in the late twentieth and early twenty-first centuries have reclaimed culture, rights, language, and land and become key leaders for a changing world. Such power from below has overthrown colonial regimes worldwide and, in the eighteenth and nineteenth centuries, ended slavery in the British Empire and the United States; Czech artists, dreamers, the Solidarity Union in Poland, churches in East Germany, and other resisters ultimately toppled half a dozen regimes in 1989; grassroots organizing has begun to change centuries of homophobia and misogyny; and the environmentalists and scientists who began to build new movements in the 1960s had to contend with a lack of even the language and concepts we now have as equipment to protect the natural world.

To hope is to accept despair as an emotion but not as an analysis. To recognize that what is unlikely is possible, just as what is likely is not inevitable. To understand that difficult is not the same as impossible. To plan and to accept that the unexpected often disrupts plans—for the better and for the worse. To know the powerful have their weaknesses, and we who are supposed to be weak have great power together, power to change the world, have done so before and will again. To know that the future will be what we make of it in the present. To know that joy can appear in the midst of crisis, and that a crisis is a crossroads.

Meteorologist and climate journalist Eric Holthaus writes, "In a climate emergency, courage is not just a choice. It's strategic. It's a survival strategy." Perhaps hope is the courage to persevere when winning looks hard; perhaps it's not hope but faith that sustains people when success looks almost inconceivable. It's in this sense that the playwright Václav Havel, who was a catalyst for revolution and regime change in Czechoslovakia in the 1970s and 1980s, speaks of it: "Hope is not the conviction that something will turn out well but the certainty that something is worth doing no matter how it turns out."

A few facts: though we are still overshooting the goals to keep global warming to 1.5°C (2.7°F), in the past two decades, climate actions and energy transitions have shifted our trajectory downward dramatically. Had there been no climate action, we might be hurtling toward 4°C or more of warming, far deeper in climate catastrophe than we are now, and far less equipped to limit it.

The movement—a shorthand for thousands of coalitions and groups and alliances, campaigns and uprisings and efforts—has done remarkable things, as my friend Renato Constantino tells in his chapter about how people from the most climate vulnerable nations shifted the agreed-upon global limit to temperature rise from 2°C to 1.5°C in 2015, through passion, brilliance, tenacity. So the movement has done a lot. Not enough yet, but a lot.

The climate movement has gone from small and polite to huge, fierce, and brilliantly strategic over the last fifteen years. It is global, and it has many victories behind it, some as tangible as a pipeline defeated, some as intangible as public awareness and wider commitment (we collected a list of them for this book; it begins on page 92). Often the changes are visible only to those who follow the policy enough to see the shift in energy sources or building codes or cutting finance for fossil-fuel projects. Sometimes victory leaves nothing to see, the trees that weren't cut down or the drilling permits that weren't issued. We need more. But we also need to recognize the victories and achievements directly behind us as well as those further in the past, because they tell us we have won, we can win, and thereby equip us to keep trying.

Climate activists have blockaded roads, pipeline paths, and fracking sites, gone on—as climate campaigner Nikayla Jefferson tells us in her chapter—hunger strikes, tree sits, marches, long walks, rallies, FridaysForFuture weekly protests; have orchestrated divestment campaigns, petitions, phone-ins, die-ins, sit-ins, educational campaigns; blockaded harbors with kayaks; hung banners from bridges; doused nude figures in fake oil in London's Tate Modern museum while it was taking money from BP; painted murals in the street; staged mock trials; shut down thousands of coal plants and prevented others from being built; stopped fossil-fuel leases and pipelines; blockaded oil trains; interrupted board meetings; organized shareholders; raised money; and, more important, raised consciousness.

Speaking of consciousness, the main job is not to convince climate deniers and the indifferent (and there are a lot fewer in either of those categories than there were a decade ago). It's to engage and inspire those who care but who don't see that they can and should have an active role in this movement, who don't see that what we do matters—that it's not too late, and we are making epic decisions now. Understanding that the consequences of activism are often indirect or slow to appear or require an informed overview—but matter nevertheless—is part of the equipment we need to see our own power and possibility.

Thanks to the divestment movement, often mocked at its inception, more than $40 trillion has been divested from fossil fuels. Public opinion has shifted dramatically, thanks to the global wake-up call of activists—and also to the increasing frequency and intensity of disasters. The majority of people do care.[2] They want to see better policies.

---

2. I write this on a day when I've just come across an article in the journal

They are willing and even anxious to see change. This popular will is also something the climate movement has built.

We have undergone an astonishing energy revolution rarely described as such, or rather we are in the early stages of such a revolution when it comes to both design and implementation. Over the past two decades, breakthroughs in technology have made renewables a genuine alternative to fossil fuel for electricity generation, one that is rapidly spreading across the world. This means that we can leave the age of fossil fuel behind. The cost of solar dropped 90 percent between 2010 and 2020, and wind is not far behind. Solar has been dubbed "the cheapest energy in history" by the International Energy Agency, which pivoted a few years ago to recognize the urgency of the energy transition and the rapid decline of fossil-fuel extraction. Every week I read about innovations and experiments in battery materials and design that mean this part of the new system is also changing, rapidly, for the better.

In 2022, investments in renewables outstripped investments in conventional energy for the first time. It is inevitable that the future will be powered largely by renewables, but the ruthlessly destructive fossil-fuel companies and their allies seek to delay the transition as long as possible for the sake of short-term profit. They can be defeated and, as Antonia Juhasz relates in her chapter in this book, have already been weakened and are struggling against their own pending demise.

Frameworks matter, too, and many of them are traps. Capitalism encourages us to imagine ourselves as consumers rather than

---

Nature Communications that states, "Americans underestimate the prevalence of support for major climate change mitigation policies and climate concern. While 66 to 80 percent of Americans support these policies, Americans estimate the prevalence to only be between 37 to 43 percent." We are the majority.

citizens; authorities like us to believe we have no power. These perspectives leave us few options but to modulate our consumption—to change nothing but ourselves and merely implore the powerful to heed our wishes. They privatize our public spirits. We need to remember our own heroic nature, our capacity for courage, compassion, and action, to remember those who came before us who took action against the odds and sometimes won. Even when they didn't, they inspired others at the time or long after to live by principle rather than by merely what is possible. Often, they changed what is possible, in part by refusing to accept what were supposed to be the limits.

People often talk about the future as if it already exists. There are parameters of what's possible, likely, and all but inevitable, and scientists have done a good job of telling us the probable consequences of what we do in the present, in terms of natural systems. But they hedge their bets because they know the future is what we make it in the present, that we're still deciding (within those parameters), because it doesn't exist yet. A lot of it is being decided now. This book is to help you recognize where the possibilities lie and what your role in them can be. We hope it reenergizes you if you're already engaged or brings you onboard if you're not. There is nothing more important in this time.

Nothing Is Inevitable

Thelma Young Lutunatabua

Climate rage smelled like pine. We were cleaning the after-cyclone refuse, and scattered branches and bits of our roof were everywhere. Often, the vestiges that don't survive a storm are the aliens and imposters—those brought to Fiji by colonizers or other outsiders. The pine trees do not have the generations of inherited knowledge that tells them how to bend with the wind and not break.

My body was still finding its breaths after the days of preparation, the hours of crouching, covering puddles, and praying that the roof remained. While outside, we heard on the radio that there was another possible storm on the way, that it could hit in the next twenty-four to forty-eight hours. Internet signal was down, so we depended on updates from a single battery-powered device. Being unable to access my usual cyclone tracking apps or even Twitter, I felt even more powerless. The vulnerability brought fear, then it brought anger.

Back-to-back Category 3 to 4 cyclones is not something that happened by chance. The intense heat wave I'm sitting in while writing this, which is making millions also seek cool shelter, is not some random occurrence.

In 2014, Évelyne Dhéliat, a French meteorologist, shared a

simulated weather forecast for what we could expect in 2050, with parts of France experiencing temperatures around 40°C (104°F). Her predictions have come true early. The mock 2050 forecast presented the exact temperatures for this weekend in June 2022.

There is a ticking clock of terror that grips so many who learn about the horrifying direction we are rushing toward. The Italian physicist Carlo Rovelli, in his book *The Order of Time*, notes that the standardization of our clocks is a modern occurrence, that traditionally time was conceived by "counting the ways in which things change." It's no wonder, then, given how swiftly the planet is shifting, that this generation is heavy with anxiety with the understanding that we are running out of days to halt the climate crisis.

If it were just a battle with time, it would be somehow easier. It's also a deep uphill confrontation with entrenched powers that have had the luxury of domination for too long. Like many others, I have joined in mass marches in New York; Washington, DC; Bonn; and elsewhere, demanding that politicians act. For decades, activists have sent petitions, held demonstrations, and faced arrest or worse. The facts have been presented and people's desires made known, yet those in power refuse to acknowledge what's truly at stake. Or they may make sweeping public statements about the gravity of the situation, but their actions speak otherwise. The latest form of climate denial is no longer refuting climate change's existence but merely recommending a slower track of action or offering ceaseless prayers to promisors of carbon sequestration. They hope that technology will prevent them from having to face the consequences of their actions.

In 2019, heads of state from across the Pacific gathered in Tuvalu, one of the low-lying atoll nations most at risk from rising sea levels. Funafuti, the capital, is not more than four hundred meters at its widest point, and its highest point is just a few meters above sea level. The welcome ceremony included Tuvaluan children sitting in a moat of water to symbolize the impending waters. The Pacific Islands Forum (PIF) meeting included Scott Morrison, Australia's prime minister, the same politician notorious for bringing a lump of coal to Australia's parliament and telling people not to be afraid. At the forum, leaders from across the Pacific implored Australia to make stronger commitments. Negotiations for the final draft of the joint communiqué were described as "fierce" and almost collapsed twice. Morrison and his delegation refused to acknowledge the role that coal was playing in furthering the erasure of island nations.

Enele Sopoaga, the former prime minister of Tuvalu, noted forcefully and tragically, "I thought, perhaps too ambitiously, that hosting the fiftieth PIF in Tuvalu, my country, which is perhaps the most threatened atoll nation due to impacts of climate change, would secure genuine sympathy and loyal understanding on our Tuvalu plea calling for urgent and complete response to climate change. Sadly, making money took precedence over saving lives in [small island developing states]."[3] Heads of state, people in power, were there face to face with Morrison, imploring Australia to care that their nations might disappear soon—and he flat out refused. Instead, he chose to perpetuate the ongoing imbalance of emissions that has led to the current injustice, with those who did the least to create the crisis bearing the worst burden.

It can be horrifically painful to know that we have the answers, to know that to halt the worst of the crisis we have to keep fossil fuels in the ground, while those in power pass permits and grant leases and make greater offers to the companies that gladly perpetuate disasters.

I wish this were just a battle against time.

For many years, there was the pervasive belief that if we could just show people that climate change was real and happening now

3. Kate Lyons, "Former Tuvalu PM Says He Was 'Stunned' by Scott Morrison's Behaviour at Pacific Islands Forum," Guardian.com, October 23, 2019.

that it would shape minds. Hence the early fascination with polar bears. But in the past just seven to ten years, we've seen a massive shift. Now, there is no doubt that climate change is real, and billions are having their lives and bodies upended because of it. In 2015, my Egyptian colleague Sarah Rifaat told me, "It's scary to think that when it becomes so extreme, and everyone feels it, will it mean it's too late?" As more and more peoples face the consequence of the climate crisis, I've thought of Sarah's question often. We need new tactics to inspire people and mobilize them to join the struggle. We need to give people better ways to get involved in the work—beyond just online sign ups, occasional marches, and more. We need better channels to put power back in people's hands. And then there's the second part of Sarah's question—is it too late?

There is no doubt that we have an immensely difficult journey ahead. It's not just an uphill climb, it is a ragged, steep mountain face we are trying to hoist billions of people up. Up to the clearing where fresh air dominates and hope permeates. While staring up at the path we must traverse, we can easily become overwhelmed and even immobilized by darkness.

Sunlight reaches only the top thousand meters in the ocean at most, yet the dark depths extend down to twenty thousand meters in many places. Ninety-seven percent of the ocean floor is in this abyssal zone no light can touch. If our planet's surface is 70 percent ocean, and if the ocean is predominantly dark, that means if someone were looking at Earth from beyond, they could perhaps register it as a mostly dark, wet, cold place. For many years, it was assumed that no life could exist without oxygen, light, or food. In 1977, in one of the most stunning scientific discoveries of the past century, scientists in the Galápagos Rift found hydrothermal vents and a whole ecosystem existing down in the depths—despite the lack of oxygen and the immense pressure. The more the deep ocean is explored, the more life-forms are being found—species with creative ways to generate their own light and showcase the multitude of ways that life can come into being.

I think of this phenomenon often, not because it terrifies me. It comforts me. If those "inhabitable" depths can be bursting with life and light, then we shouldn't be a planet afraid of darkness.

Nothing is inevitable, and that's crucial to remember in this fight. It's no surprise that so many of the tactics the climate movement is using to tackle fossil-fuel Goliaths are taken straight from playbooks used to bring down dictators. If we can get at the fossil-fuel industry's pillars of support, the legs that prop them up, they will tumble. The Serbian activists who helped overthrow Slobodan Milošević have been spreading this methodology for years with the simple truth, "If people withdraw their support, the ruler cannot rule!"

There is immense momentum now in the climate movement, with coal, oil, and gas projects struggling to find funding or insurance backers. Climate groups have been systematically targeting the monetary pillar of the fossil-fuel industry and finding success. There are those working to remove the political power of those who are in the pocket of coal, oil, and gas. One piece of good news is that, in the May 2022 Australian elections, Scott Morrison and all his coal-loving friends lost their seats in government, largely due to waves of organized people frustrated with climate inaction.

Sitting in the crevices of the hearts of many, even after a victory, are the questions, *Will it be enough? Will my actions be enough to keep us from catastrophe?* We can track the global carbon emissions, monitor the temperature levels, keep an eye on the receding glaciers—and then use these as measuring sticks for our own behavior. However, staying in this state of berating urgency can also be dangerous. The question shouldn't be *Will my actions be enough?* but *Will* our *actions be enough?* This is a communal quest in which everyone can bring their talents, visions, desires, access—and if one person struggles, we can help each other up.

There is a gravity to the moment we are in, a horrid perpetuating whirlwind that is coming. There is no doubt we have to act fast to transition the world's energy. Billions of lives are at stake. Yet, within that space, we also need to keep room for the joy and beauty of this world to take root. We also need to ensure we remember our shared humanity and the intersectionality of all our struggles. In our efforts to protect what is being lost, how do we make sure we're appreciating what is fighting to remain? After a frustrating new law is passed, do we take a moment to bring bread to our neighbors? Do we watch the sunset after a brutal storm? Have we acted as if there were possibilities other than destruction? Have we taken steps to show that our liberation is tied to that of others?

I've seen too many activists hold the heaviness of the climate clock within them, a well-meaning martyr complex that prevents any intake of abundance. I've also seen activists push aside any injustice that wasn't directly related to limiting carbon emissions. There is a tender balance between urgency and humanity, and the truth is that keeping our brilliant selves whole and standing in solidarity with others will deliver so much more in the long run. The poet and noted ornithologist J. Drew Lanham aptly said, "Joy is the justice we give ourselves."

We cannot honestly know if our actions will be enough, but that doesn't mean we can't try. What we hope to do with this book is to provide that light in the dark. When the pressure is pounding down on you, this anthology is the bioluminescence that can be a glimmer of encouragement. If you are terrified, that could be a good thing, because it means you're paying attention. What you can't do is let it immobilize you, to pull you down to worse depths. We have to learn how to bend with the intense winds and not be so rigid we break like the pine trees.

WE HAVE THE SOLUTIONS

There is too much bad news to justify complacency. There is too much good news to justify despair.

—**Donella Meadows**

The ultimate, hidden truth of the world, is that it is something that we make, and could just as easily make differently.

—**David Graeber**

Not everything that is faced can be changed, but nothing can be changed until it is faced.

—**James Baldwin**

This seems an age of catastrophe, but it's also an age equipped, in an abstract sense, with all the tools it needs. Utopia is available to us. If, like me, you lived through the civil-rights movement, the antiwar movement, you can never discard hope. I've seen social miracles in my life, ones that have stunned me—the courageousness of ordinary people in a struggle.

—**Mike Davis**

Here's Where You Come In

Mary Annaïse Heglar

It was December 2019. Before COVID-19 had become a pandemic. Before we even knew who would run in the 2020 presidential election, much less who would rouse his supporters to storm the Capitol. Before I thanked New York City for the fifteen years it gave me, packed my things, and moved to New Orleans.

The calendar said it was December, but the temperature felt like March. I took the 4 train from my home in the South Bronx to an apartment near Union Square. I had been invited to lead a conversation with a handful of artists about the climate crisis and their place in it. I didn't bring a coat.

What I found was an intimate gathering of six or seven people. As we milled about over plates of cake and mugs of coffee, we started sharing our stories about Hurricane Sandy. We marveled at how much our experiences differed based on our neighborhood. Sandy turned the Lower East Side—which was originally built for low-income communities but is now fairly affluent—into a place with flooded police cars and no electricity. Meanwhile, the South Bronx, originally built for the affluent but now the poorest congressional district in the country, came out of the storm relatively unscathed, thanks to its position on higher ground and its physical connection to the mainland of New York State.

We speculated about how long before another Sandy—or something much, much worse, perhaps something we don't even have language for yet—pushed the masses from Lower Manhattan into the South Bronx. Then where would my neighbors go?

From there, the conversation naturally spiraled into the undercurrent of terror that comes with being alive today. At the time, Australia was ablaze, and the embers had barely cooled in the Amazon. A typhoon was encroaching on the Philippines. And that wasn't counting the countless other disasters underway in Africa and Latin America that never made the headlines.

This was before the devastating 2020 hurricane season that exhausted the conventional naming convention in such a way that it was permanently retired. This was before the term "fire-nado" was common. This was before San Francisco turned an eerie orange. So many precedents had not yet been set. But still: climate change was in the air and, thus, on our tongues.

I could tell that it felt good to talk like this: open and honest about the experience of watching the world fall apart in front of our eyes. To say our fears out loud and have them, and ourselves, accepted and understood.

I could see the weight lifting from our shoulders, but it only rose so far. It hung in the air, just above our heads like a heavy ominous cloud, until someone popped the question that brought it right back down on us: *But what can we, as individuals, do?*

In the years since 2018, something remarkable has happened to the climate conversation. It finally found its way out of the academy, oozed out of the Big Green groups and expert circles, and landed in the streets and on everyone's lips. Now, I hear it everywhere: on the street, in the subway, in the airport, in the changing room at my yoga studio, in the checkout line at the grocery store. It's no longer something frontline communities discuss only among themselves. While they might be the ones feeling the pain, they're no longer the only ones afraid of it. Climate is no longer niche. It's mainstream.

It's about time.

There's a lot of people to thank for this blessed change. The 2018 Intergovernmental Panel on Climate Change (IPCC) report, which used some of the starkest language the scientific community has ever used, woke a lot of people up. It spelled out in brutal and unequivocal—but, most important, *honest*—terms the consequences of a runaway addiction to fossil fuels. Then there was the international climate youth movement that made it clear that you don't have to have a PhD to understand climate change—you don't even need a high school diploma. Then, there are the frontline communities that have been stomping and screaming about the crisis on the horizon for decades. For them, it was never on the horizon; it was on their backs. There are so many people to thank.

For me, as overdue as this shift might be, it's also bewildering. I am what the meteorologist and journalist Eric Holthaus calls a "Climate Person"—someone whose whole life is bound up in confronting the reality of the climate crisis. I joined the environmental movement in earnest in 2014, when I began working for one of the biggest green groups in the country. In 2018, I began speaking out on my own—in essays, on panels, and in Twitter rants. This made me not just a Climate Person but a Public Climate Person.

In 2019, we Climate People were used to being a small group. Marked by our intimacy with one another, our knowing glances across rooms. We were used to being mocked and sidelined as the killjoys, the bummers. In public places, we intuitively gravitated toward one another, carving out our own little corner of the party or our own sliver of the internet known as #ClimateTwitter, where we could rant and rave and scream and grieve together. And get on each others' last nerve.

But our cover has been blown now, and the doors of our clubhouse have been torn off the hinges by hordes and hordes of brand-new Climate People. It seems every day there's a new wave of people who have woken up to the awful, awful truth that this crisis is bad, it is getting worse, and no one is coming to save us.

Once you see it, as we Climate Folk well know, you can't unsee it. And when the shock finally passes and you find your feet again, you're overcome with the urge to do something—anything—to wash as much blood off your hands as possible.

Suddenly, Climate People are popular! Where we used to quietly lament our lack of dinner party invitations and hold our own parties in secret, we're now the belles of the ball. Before, people rolled their eyes, smacked their teeth, and backed away when I mentioned my work. Now they lean in close. They ask questions and actually listen to my full, uninterrupted answer—men included!

And no question is more fervent, more persistent, more desperate than the one that weighed us all down that December evening: *But what can I do?*

There's probably no question that Climate People hear more, and fear more, than those five words. The askers get more and more frustrated, their newfound sense of urgency threatening to burn a hole in their throats.

They know it's about more than recycling, "buying green," and turning the lights off when they leave the room. They've gotten the memo that we need structural change in addition to individual change. They've processed past the shock. They're ready to get to work. Why, they demand to know, can't I give a simple answer to such a simple question?

Here's why: Because if you want a real answer—one that won't leave you with tiny solutions that will ultimately disempower you and burn you out—you have to understand that the question is profoundly complicated.

BELIEVE ME, I understand why that question seems so cut and dried. But that's just an illusion conjured by several fallacies. And perhaps the first thing a new Climate Person can do is understand them.

Let's start with the first fallacy: that climate action comes in individual packages. Almost every time I hear people struggle to find their place in the climate movement, it's because they feel too small to make a difference. They know that the world needs to essentially bring fossil-fuel production to a screeching halt, not just now but RIGHT NOW. And they know that no single action they take can bring that about. So then what?

Well, what if your power in this fight lies not in what you can do as an individual but in your ability to be part of a collective? What if you broadened your perspective beyond what you can accomplish alone and let yourself see what you could do if you lent your efforts to something bigger? Yes, it's true that you can't solve the climate crisis alone, but it's even more true that we can't solve it without you. It's a team sport.

There's another fallacy embedded there: the expectation that a single, neat behavioral change will be enough. I've done a lot of interviews, sat on a lot of panels, and I've often heard the question *What can I do?* boiled into an even more maddeningly and damningly simplistic form: *What's the* one *thing people can do?* There's no such thing. I wish there were.

Especially now, at this critical stage, we have to accept that we're all going to have to buckle down for the long haul. Responding to this crisis is going to have to become part of who we are. All the time. Once you understand that, you understand that this isn't about climate *action* at all. It's about climate *commitment*. Climate action is recycling or voting or opting for a vegan meal. Climate commitment includes those singular actions, but is bigger still. It's a framework. It's asking yourself: *What can I do next?* And always next.

Then there's that other alluring fallacy: the idea that if we do the right thing, we can put an end to this madness. That there's a stop button somewhere.

As the climate scientist and brilliant writer Kate Marvel puts it, "Climate change isn't a cliff we fall off but a slope we slide down." The climate has already changed. If you're reading this, for all intents and purposes, you are doing so on a planet that is fundamentally different from the one you were born on. And what's been done, sadly, cannot be undone, at least not in the near future. But there's real good to be done by not letting it get worse. Limiting the damage is good, noble—valorous even. Any suffering we can alleviate is a blessing.

By now, you're probably becoming either consciously or subconsciously aware of the heartbreaking truth at the core of the climate crisis: it's so unfair. It is.

That's probably the simplest thing about climate change—the injustice. It's apparent at the macro and micro scales. The parts of the world that contributed the least to the crisis will suffer first and worst. Mere children have been thrust into positions in which they have no choice but to fight for their lives, for their right to see the stable planet they were taught about in storybooks and science books but have never seen in real life.

To call it a crime against humanity would be an understatement.

But now that you're aware of that truth, it's crucial to remember one thing: it's not enough to be right. The facts have been on our side for a very long time, but we're still losing. Why? Because this isn't a spelling bee or a standardized test. This is a fight for justice.

The climate crisis is, in more ways than I can count, the ultimate culmination of a centuries-long run of exploitation and extraction, including slavery and colonialism and all of their offshoots. Those horrors were all justified by some measure of pseudoscience that could have been—and was—easily disproved. But that wasn't enough. So it is with the climate crisis.

The scientists and experts have studied the problem and the solutions and presented their findings ad nauseam. But it wasn't enough. Because this isn't just about science or facts. This is about power. And it's going to take an army. That's where you come in, new Climate Person.

I know it might not sound like it, but there's a lot of good news in there. For one thing, you don't have to do this all alone. In fact, you can't. Check your savior complex at the door, please. Because we're talking climate commitment and not a single climate action, that means you don't have to get it exactly right every time. This is a practice, not a perfect. The fact that every fraction of a degree of warming—Celsius or Fahrenheit—matters means that you're never too late or too small to help.

The right time to start your climate commitment is always right now.

But the question remains. *What can I do?* Well, now that you understand that the question is complicated, the answer actually emerges as quite simple: do what you're good at. And do your best.

If you're good at making noise, make all the noise you can. Go to climate strikes, call your representatives, organize your neighbors. Vote. Every chance you get. If you've got it in you, run for office or volunteer for a campaign. Join something bigger than yourself because this is so much bigger than any of us alone. It's about all of us, together.

If you're raising children (and they do not have to be your children—nieces, nephews, and play cousins all count!), teach them to love the Earth and to love each other, teach them the resilience that shows up as empathy. If you're good at taking care of people, take care of the legions of weary climate warriors. If you're a good cook, cook. Find the people helping and help them. Make it as sustainable as you can within your means, but more than anything, share it, build a community around it.

The artists I spoke to that balmy December evening lamented the fact that they weren't engineers or scientists or some other type of "expert." But, as I told them, it is not their job to design the policy plans for rapid decarbonization, to decide which coal plants to shut down first, and what exactly to replace them with. We have people on that. As the writer Toni Cade Bambara once put it, the role of the artist is to "make revolution irresistible."

Make no mistake about it: overthrowing the fossil-fuel industry is nothing short of a revolution, a rebirth. Our whole world was built on fossil fuels. It never should have happened, but it did. That's why we need a whole new world, and we all, every single one of us, has a powerful role to play as a midwife in this rebirth.

Viewing through that lens, you begin to see that you're not powerless at all. Far from it. Someone built this world, with all its inadequacies and inequities. And, yes, that world is falling apart. But, if someone could build that world, I have to believe we can build a better one. We can sort through the rubble of the old world, keep the things that serve us, and bury the things that don't. We can start fresh, knowing everything we know now. We can be stronger, kinder, braver. How beautiful would that be?

That's why it's so impossible for any Climate Person to tell any other Climate Person, new or old, what their own climate commitment should look like. We don't know what special thing you bring to the movement—only you know that. And we can't wait to see the magic that will happen now that you're part of our world.

We Are Not Doomed to Climate Chaos

Edward R. Carr

Anyone following discussions of the February 2022 report of the United Nations' Intergovernmental Panel on Climate Change (IPCC) could be forgiven for believing that we are doomed to climate chaos.

But, in fact, this report did not say that. I know because I was a lead author. We found that we can get on clear pathways to a climate-resilient future for everyone by connecting actions that reduce greenhouse gas emissions (called climate change mitigation), manage the impacts of climate change we can no longer avoid (climate change adaptation), and promote a sustainable future of decent living standards and opportunity for all (sustainable development). There is substantial evidence that changing the politics and norms around who participates in climate action enables such connections, and working on these challenges at the local level produces solutions tailored to specific opportunities and needs that can aggregate up to global impacts.

The report makes it clear that we have waited too long for incremental changes to effectively bring us to a climate-resilient future.

This is not a message of catastrophe, but it does demand we think transformationally. We live within large structures, from our transportation networks to our massive, highly integrated global food system. Dependent as we are on these structures for our well-being, individuals have a limited ability to transform how they live on their own—a single consumer's purchasing decisions will not shift the cultivation techniques of a global agribusiness. While we all have a role to play in addressing climate change, that role must include, and perhaps even emphasize, transforming these structures.

There are clear levers for change, even at the local level. The IPCC report presents evidence that one concrete path toward structural transformation lies in rethinking how we approach climate change mitigation, climate change adaptation, and the achievement of sustainable development. These three arenas have long been independent silos of knowledge and action, which a growing body of evidence indicates results in the greater likelihood of unwelcome trade-offs and compromises. For instance, efforts to manage sea level rise with a seawall can produce real adaptation benefits for those behind the wall. If that seawall displaces flooding and erosion onto vulnerable communities that cannot afford similar construction, it will compromise their pathways to sustainable development.

Should the Massachusetts Decarbonization Roadmap result in climate change mitigation policies that emphasize disincentivizing driving without also enhancing regional public transportation options and remote work options, the state risks encouraging a surge of people moving into Boston and its coastal suburbs for transportation options (but which would also put them closer to coastal flooding). Increasing the population density in the areas most challenged by sea level rise is not only potentially maladaptive, it can also result in increased rents and property prices that limit the ability of vulnerable and marginal communities to build sustainable lives in these coastal cities and towns.

The IPCC report found that effectively navigating the trade-offs and synergies between mitigation, adaptation, and sustainable development requires much greater attention to whose lives, livelihoods, and development we are trading off or supporting. At one level, what communities, states, and countries identify as a problem and choose as solutions depends on who is in the conversation. Attention to previously marginalized voices can bring forth new opportunities for synergies among mitigation, adaptation, and sustainable development efforts while revealing previously hidden or overlooked trade-offs and impacts. Broader participation in climate change planning and action will yield more inclusive and effective pathways toward resiliency.

At another level, the IPCC report tells us that transformation is now inevitable. The choice is between the transformations we choose and those forced on us by the climate we have altered. We will have to make decisions about what we transform, who reaps the benefits of that transformation, and who bears the future burdens. If these are to be decisions that produce both just and durable improvements to human well-being, then we will need to open up the political processes through which they are made. The processes that matter are not just global or national in scope. Effective climate action speaks to local circumstances—the local environment, economy, and society. Our climate future will emerge through local politics' operating in the context of national policies and global initiatives.

Where once we emphasized science and technology as the bridges to a climate-resilient future, the IPCC report recognizes a fundamental fact: climate change is a problem of people. We already have the technology we need to mitigate and adapt to climate change in a manner that brings about sustainable development. We have failed to act on it.

Our message of transformation is a message of urgency and hope. A climate-resilient future is still possible. It starts with changing who is in the conversation to better identify the challenges we need to address, the best ways to address them, and the just, equitable outcomes we expect from our actions. From these conversations will flow clear rationales for what we need to transform, what we want to keep, and why. This, in turn, will point us to the linked adaptation, mitigation, and sustainable development actions that will work to benefit the widest set of people in communities, states, and countries.

Defeating the Fossil-Fuel Industry

A Conversation with Oil Policy Analyst and Investigative Journalist Antonia Juhasz, Rebecca Solnit, and Thelma Young Lutunatabua

Rebecca: You recently called fossil fuels "obsolete," declaring that "the global oil industry is in a tailspin" and "the end of oil is near." In what ways is that true, and how do we make it even more true?

Antonia: The oil industry in particular, and the fossil-fuel industry more broadly, has been suffering death by a thousand cuts for years. Until very recently, corporate profits, market values, investor returns, and demand growth had been in a steady nosedive. By 2019, the fossil-fuel industry ranked dead last among major investment sectors in the United States. The COVID-19 pandemic exposed all of the industry's frailties and made each of them worse. When Russia went to war against Ukraine, the global community's response demanding an end to fossil fuels was unlike any I'd ever seen.

The weakening of the industry is due to concerns about climate, concerns about public health, equity, the corrosion of politics, the tendency toward war, the tendency toward pollution and corrosion that the industry has caused. There has been a steady movement organizing against it, and that has led to policies being put in place across the world that have been geared toward ending the use of fossil fuels. Those have come in the form of pressing for expenditures on renewable forms of energy and healthy transportation instead of using fossil fuels; putting in place much tighter regulations that hold public health, well-being, and justice in much higher regard; trying to curb the political and financial influence of the industry. Confronting the climate crisis in myriad ways. Trying to replace extractivist with regenerative economies.

As the climate crisis has worsened, more and more people have experienced being on the front lines of climate harm, and they have also increasingly come to know that the primary cause of the climate crisis is fossil fuels. Fossil fuels, hands down, are the main contributor to global warming, and if you want to address the climate crisis, you have to address fossil fuels. The more the climate crisis worsened, the more that people were forced to acknowledge that reality over and over again. If you want to confront the climate crisis, you have to confront fossil fuels.

As more people confronted the crisis and wanted to find solutions, they began to realize that there have been people on the front lines of fossil fuels, of climate crises ever since the industry began. Increasingly, people came to look to those who had the longest experience confronting the fossil-fuel industry. Those were primarily marginalized communities, communities of color, nations of color, who have been since the beginning on the front lines of fighting fossil-fuel extraction, who were the most immediately harmed by climate change, and who have always been resisting the industry. What if all of those strategies that were already being utilized and practiced were supported, elevated, echoed? What if those leaders were given the resources they needed, the backing? Increasing support of those who have been on the front lines presenting solutions to fossil fuels and the climate crisis has had a dramatic impact.

As all of that has expanded, the pressure on the fossil-fuel industry has expanded as well. The acknowledgment that people, environments, communities are impacted at the site of exploration, at production, at transport, via consumption and waste; that the industry wields its influence from local communities to international arenas; and that all of those can be points of resistance as well, has grown and blossomed. The industry is being confronted more and more at every one of those junctures, both physical as well as where it wields financial influence. Each one of these junctures has seen attempts by activists, by policymakers, by the media, to increasingly hold the industry to account and restrict its influence and power.

At the same time, the industry has pushed back. There are many nations whose economy and power reside entirely with the fossil fuel-industry. Those nations and those industry sectors that are the most powerful have also doubled down on their intent to stay in power and relevant.

Rebecca: I've heard it said that the fossil-fuel industry will inevitably end, and what we're really deciding now is whether it will be a managed decline—how fast it will end.

Antonia: Yes. I think the age of oil will come to an end, and it will either come to an end as the human species is so overwhelmed— just focused on getting through one crisis or another—that we can no longer produce or consume oil. Or it will end because of a managed decline—the intentional policies that make it possible to move as quickly as possible off of fossil fuels, with policies that we already know that will make our lives healthier, more sustainable, more equitable, more just. We rapidly implement those policies and bring fossil fuels to an end. Or we let fossil fuels end us.

Thelma: How do we have this transition in a way that doesn't leave everyday workers in a bind? How do we help the workers, whether they're in Saudi Arabia or in Louisiana? How do we help them transition as well?

Antonia: I think one of the most important things to acknowledge is that the fossil-fuel industry has been successfully and intentionally shedding jobs for decades. The industry wants to shed one of the most expensive parts of the industry, which is workers, and it has been doing that. It's shrunk by about 150,000 workers in just the last several years. Seeing the fossil-fuel industry as a secure place for workers is false, a myth that really shouldn't be perpetuated. It has never been safe, has never been healthy. With a few exceptions, it's been a sector that has fought unionization, has fought creating a safe and healthy environment for its workers, and has fought attempts to provide, for example, long-term health benefits and security for its workers. That said, there are still, of course, many communities where workers are dependent on fossil fuels. Studies have made clear that the more rapid the transition, the less harm will be experienced. Delaying and slowing are making matters far worse, as the repercussions of the war in Ukraine have made painfully clear. We are, however, already seeing in the US a lot of policies focused on the transition. International labor organizations and climate justice movements have created clear policy paths and recommendations focused on ensuring that those communities that have been most harmed by fossil fuels are given the greatest means and support to transition. And on ensuring that the perpetrators of the climate crisis provide the financing and mechanisms that can enable and speed this transition.

Rebecca: Does the fossil-fuel industry acknowledge its own demise? What signs of that do you see?

Antonia: The fossil-fuel industry knows that it's in trouble, and it's been trying to reshape itself since the pandemic. The form that it's been reshaping itself into has varied. One of the things the industry has been trying to do is consolidate itself into the places that are the cheapest and easiest to produce. Another approach has been to acknowledge the reality of the climate crisis, the necessity of reducing emissions—but then saying that the responsibility is with the consumer, as if to say, "Look, it's not our fault, it's yours." Believe me, they intend to keep producing as long as they can. They have no intention of going anywhere, and they also have no intention of becoming renewable energy companies—nor do we want them to. The direction we want them to move in is whittling down aggressively through managed decline.

Thelma: My question for you, then, is, what could the future of energy look like? And what role can people play in facilitating that managed decline, in helping create the new energy options?

Antonia: The first thing to say is that the fossil-fuel industry does not always win. When we think about the energy future, we can just look at the present. The truth is, fossil-fuel consumption is really driven by a very small number of people. The wealthiest people who live in North America and Europe are, hands down, the primary and most destructive consumers of fossil fuels. If the wealthiest 10 percent of the population continued to use energy the way it currently does, and the entire rest of the world went to net zero tomorrow, we would still not be able to reach our climate goals. The vast majority of the world is already consuming energy just fine, particularly in Indigenous communities and in communities outside of the US.

We also want to ensure energy justice; we don't want to see energy poverty. One of the best ways to expand energy justice is to follow models that already exist, where you have a far more localized, democratized provision of energy, where you have the shortest distance possible between the energy that you produce and the energy that you consume, and you have, as much as possible, a democratized, cooperative model of funding and supporting that energy. The most important thing about our energy future is to embrace all the powerful solutions that already exist in our energy present, and to know that we do not need fossil fuels to meet our energy needs.

Rebecca: What would be your call to action?

Antonia: So many incredible movements have had already such a profound impact on the fossil-fuel industry. There are so many places to join, whether it's supporting frontline communities and their resistance efforts or working on divestment campaigns that are trying to get banks to stop investing in fossil fuels. There are so many political candidates who are running on platforms focused on ending the influence of the fossil-fuel industry—local campaigns, state campaigns, federal campaigns. There are lawsuits that are winning and challenging the industry at every turn. There are efforts at the United Nations that have just succeeded in declaring it a human right to have a healthy, safe, and clean environment.

There are protests taking place every day. The challenge is simply to be able to say, *We know that fossil fuels are the central cause of this problem, and we know that we have solutions.*

A Climate Scientist's Take on Hope

Joëlle Gergis

As a climate scientist, people often ask me what is the single most important thing they can do to address the climate crisis? My answer is simply this: recognize that you are living through the most profound moment in human history. Averting planetary disaster is up to the people alive right now. When you realize that the 2020s will be remembered as the decade that determined the fate of humanity, you will tap into an eternal evolutionary force that has transformed the world time and time again. Recognizing you are part of a timeless tug of war for social justice electrifies the present moment in a way that brings meaning and purpose to our lives.

We have urgent decisions to make about how much will be lost to future generations.

We must choose what we are willing to save.

In my work as a lead author on the latest Intergovernmental Panel on Climate Change (IPCC) report from the United Nations,[4] I came

---

4. See also Edward Carr's reflections on the IPCC report starting on page 29.

to understand that the only way out of the mess we are in is first to acknowledge the enormity of the crisis we are facing. Every single decision humanity has ever made to exploit nature and the world's poor has culminated in the planetary destabilization we are currently experiencing. Humans have altered every component of the Earth system, from deep ocean trenches to the highest mountaintops. Around three quarters of the land surface has been ravaged by human activity; only 3 percent of the Earth's land ecosystems are ecologically intact. Almost half the carbon dioxide accumulated in the atmosphere has been emitted since 1990, overwhelmingly by rich countries. The horrifying scale of destruction and entrenched inequality has led us to a place where we need to redefine our shared values of what it means to be human at this moment in time. Are we all okay with the fact that around a quarter of the world's urban population lives in slum conditions, or that 90 percent of seabirds alive today contain plastic in their guts? Are we really going to sit back, watch, and declare it all too late, that there is nothing worth saving? Is this really the best we can do?

The truth is that some of the changes we have now unleashed in frozen places and oceans are irreversible and will be with us for centuries. We have set in motion the melting of mountain glaciers, permafrost thaw, ocean acidification, and sea level rise that will be with us long after we reduce greenhouse gas emissions. It's an appalling reality to come to terms with—the Earth's equilibrium is now shifting in a fossil fuel–cooked world. But just because the situation is dire doesn't mean it's too late to try to stabilize things. Perhaps the most important message to come out of the latest IPCC report is this: *How bad we let things get is still in our hands.* Additional warming will be determined by future emissions. That is, what policies governments around the world choose to do to reduce greenhouse gas emissions and restore ecosystems will directly lead to restabilizing the Earth's climate. Or not.

The good news is that there is no evidence to support the notion that we are currently facing runaway climate change or the inevitability of an unlivable future. Once emissions start to stabilize, temperature follows suit. Sometimes this message is misunderstood by doomers and nonexpert commentators, so it is worth being clear about what the science has to say. The IPCC report explains that every single metric ton of carbon dioxide we prevent from entering the atmosphere lessens the severity of the impacts we bake into the system. Our assessment meticulously describes how every fraction of a degree of warming matters—the scale and severity of impacts begin to compound and cascade with higher levels of warming.

Historical emissions have warmed the planet by approximately 1.2°C above preindustrial levels, and we are on track to breach 1.5°C in the early 2030s under all emission scenarios laid out in the IPCC report. The likelihood of reaching 2°C occurs as early as the 2040s under high emission pathways, something we still have the power to avoid. If we choose to drastically reduce emissions, we could see a temporary overshoot of the Paris Agreement targets, then a restabilization of temperature as emissions begin to fall. To use a medical analogy, if we choose to treat our cancer early enough we can stop its spread, but the longer we delay, the more limited our treatment options and the worse the damage becomes.

Right now, global emissions pledges under the Paris Agreement are not enough to keep warming under 2°C. Currently implemented policies have us tracking 1.9° to 3.7°C of warming by the end of the century, with a best estimate of 2.6°C. If countries fully implement the long-term net-zero emissions targets promised at the UN climate summit in Glasgow in 2021, then we are likely to see global warming stabilize between 1.4° and 2.8°C by 2100, with 2°C considered a most likely "best case scenario." The reality we have to face here is that our collective action is still not enough to stop major components of the Earth's biosphere collapsing or avoid displacing millions of people from rising seas. That simply means we need to redouble our efforts, not give up.

While there is limited evidence of abrupt climate change on a global scale under 2°C of global warming, the probability of crossing thresholds that activate tipping points like destabilizing ice sheets and the abrupt thawing of permafrost increases with higher levels of warming. Although there is some evidence to suggest that some tipping points like the stability of the Greenland and West Antarctic ice sheets may be triggered in the Paris Agreement range between 1.5° and 2°C of global warming, the risk is much higher between the 2° to 3°C of warming expected on current policy trajectories.

The IPCC report says that unless we cut emissions in half by 2030 and reach net-zero emissions no later than 2050, we are on track for catastrophic levels of warming that will profoundly alter all life on Earth. At sustained warming between 2° to 3°C, the Greenland and West Antarctica ice sheets could be lost almost completely and irreversibly over thousands of years, committing us to between two and ten meters of sea level rise over the next two thousand years. Currently, around a quarter of a billion people live on land less than two meters above sea level. The latest IPCC assessment estimates that approximately one billion people living in low-lying areas are projected to be at risk from sea level rise and storm surges as early as 2050. This is still something we can avoid if we relentlessly pressure our leaders to do better.

As we have already seen, 1.2°C of warming has caused dangerous and widespread disruption to nature and human societies. Imagine the impacts that will unfold with 1.5°C, 2°C, or even 4°C of global warming. We are currently failing to outpace escalating and compounding risks in many parts of the world, particularly in

Africa, South Asia, Central and South America, the Arctic, and small island nations. The situation progressively worsens once we reach 1.5°C, when the adaptation limits of many ecosystems are reached. Between half and two-thirds of all species across terrestrial, freshwater, and marine systems have already shifted their geographic ranges in response to the global warming we've experienced so far. Beyond 2°C, the IPCC says that adaptation is simply not possible in some low-lying coastal cities, small islands, deserts, mountains, and polar regions of the world.

The scientific community is saying that we can still avert the worst aspects of climate change, but we need to move quickly. We must respond as we would in an emergency, as in the COVID-19 pandemic. What we do over this coming decade is literally a matter of life and death. We must decide that destabilizing our planetary home is simply not an option.

Which brings us back to the realization that people alive today will determine humanity's future—averting planetary disaster is up to us. The idea that the climate change problem is fundamentally unsolvable is not only scientifically wrong and unhelpful, it also tragically leads people down the same path of disengagement and inaction that allows fossil fuel burning to continue. We will not see the political response we need to address climate change until we redefine the cultural and social norms that are destroying life on Earth. As voters and consumers, we are responsible for creating or removing the social license needed to maintain the status quo of burning fossil fuels and the destruction of nature to the point of planetary instability. Every decision we make can be a decision to stop trashing the planet.

As a scientist, I don't believe that our messages are ignored by the public because people don't care. Many people don't have a science background or find it challenging to stay engaged in a technical debate that can be really alienating. But, as worsening wildfires, drought, floods, and storms continue to assault our world with every passing season, people are waking up to the fact that the reality of climate change is being experienced right now, whether we "believe" in the science or not.

So, how do we live in a moment as fraught at this? How do we continue on when the enormity of the challenge sometimes feels so insurmountable, so futile?

What gives me hope is that human history is full of examples of people across the ages who have risen to face the great challenges of their time and succeeded against all odds. Victory is not the arrival in some promised land; it is a series of imperfect victories along the way that edge us closer to building the critical mass that eventually shifts the status quo. Right now, we are living through the biggest social movement of our time. A time of true global citizenry, driven by our duty to protect the planetary conditions that sustain us all. The question is, *How are you going to show up in this moment?*

Recently I was speaking about climate change at a bar in a small town in rural Australia, where education is low and skepticism is high. I spoke to the gathered crowd about the escalating impacts of a warming world on our magnificent country. Here in Australia, we are still dealing with the aftermath of the 2019–2020 wildfire season— the most catastrophic in our history—when a quarter of temperate forests went up in smoke, killing or displacing three billion animals. The devastation was so great that the koala—Australia's most iconic animal—is now listed as an endangered species along our east coast. In the handful of years since 2016, we have lost at least 50 percent of the Great Barrier Reef. The largest living organism on the planet is now dying. We know in our bones that the world has changed. We can see it happening outside our window, and yet we still have a government that plans to support the continued extraction of fossil fuels because we let them.

As I was getting ready to leave the crowded pub that night, a skinny man with a wrinkled face carefully approached me. He waited for the swarm of guests to dissipate before stepping forward to explain that he was just walking past the bar that evening and saw that a climate scientist was speaking. As a country bloke who lives on the land, he'd never been a "believer" in climate change. But after hearing me speak in simple terms, an epiphany shot through him. The error of his ways sickened him.

With red-rimmed eyes and a wavering voice, he said, "I want you to know that I am sorry. I'm sorry it's taken me so long to understand this." I held the gaze of a man softened by humility, and gently said, "It's all right. You are here now. It's not too late."

Change happens gradually, then suddenly. It's never too late to be part of the social movement that will help heal our world.

From Destruction to Abundance

Leah Cardamore Stokes

Fossil-fuel companies and electric utilities spent decades telling the world that climate change wasn't real. "It's a hoax," they cried out, in newspaper ads, glossy reports, and high school textbooks. Their propaganda was pervasive—and effective. Throughout the 1990s and 2000s, the exact language from climate denial documents ended up in the news media and even American presidents' speeches.

When denial became indefensible—in the wake of another "thousand-year" drought, or one more "never-seen-before" heat wave—the fossil fuel industry started singing a new song: the crisis can't be solved. "There are no alternatives to fossil fuels," they claimed. "Maybe later we could figure it out, but for now, we must stay the course." Delay paid them in cash. With every extra day, these companies could steal a bit more money from the ground. And, in the process, steal a bit more from our future.

Before long, another song rose up to meet all this denial and delay: "We have the solutions now!" This music came from scientists and from the people. It told us that fossil fuels are a poison and that clean electricity and electrification are the cure. This message was not broadcast as far or as wide because it didn't have the same money backing it. But it had the advantage of being true. And that fact alone allowed it to spread.

Pollution touches every part of our modern lives. But it doesn't have to be that way. Think of all the planet-warming pollution that the United States pumps out each year. It's a vast amount of largely invisible molecules that come from machines as varied as pipelines, power plants, cars, stoves, and lawnmowers. If we gather all that pollution, we could divide it into chunks, like a pie. Each slice of that pie is a piece of the economy that we can move away from dirty fossil fuels and toward clean technologies.

We can eliminate the first quarter of our pollution by cleaning up our electricity system. Right now, more than half of America's power comes from burning coal, fossil gas, and a bit of oil. Globally, around two-thirds of electricity is generated by fossil fuels. But we do not have to keep digging up ancient flora and fauna to make electricity: we can use the wind, the water, and the sun. If we consider all the power built in 2012, around half of it uses clean energy. Fast-forward a decade, and, by 2021, more than 85 percent of the new power built that year runs on renewables. That's real progress. We should celebrate it and keep pushing. We can't rest until every dirty plant is shut down and every new power plant is a clean one. And we must work to shut down all the existing dirty power. When we get 100 percent of our electricity from clean energy sources that will be a happy day.

The wires that carry all this power are infinite tentacles, reaching into every nook and cranny of our human world. For most people, electricity is there when they need it. And that means it can be a catalyst for cleaning up another quarter of our pollution problem through electrification. Right now, in the United States alone, a billion machines run on dirty fossil fuels. We can and must electrify all these machines. To do that, we need to make progress on cleaning up our transportation system, buildings, and parts of heavy industry.

When we move things around—whether it's people, food, or machinery—we use energy. Right now, transportation relies on fossil fuels. But we can change that. Rather than filling our cars with gasoline, we can fill them with electrons charged from the sun. Our school buses and streetcars and subways can all run on electricity. Even bikes can now use electricity, making climbing a steep hill a breeze. While we can't electrify every part of our transportation system with the tools we have today, we can make a massive amount of progress. And when we electrify all the trains and mini-planes and automobiles we can, we will eliminate that second quarter of the pollution problem.

The third quarter of the solution is found in electrifying two more sectors: buildings and industry. Right now, many of us are effectively running mini fossil-fuel plants from our homes. That's one way to see the gas furnace in the basement and the gas stove taking center stage in the kitchen. You probably wouldn't feel good about living next to a fossil-fuel power plant, so why do we invite these dirty appliances into our homes? Heat pumps and induction stoves can eliminate that pollution. A heat pump is an efficient electric machine that both heats and cools your home—no fossil fuels necessary. An induction stove uses the power of magnets to cook food fast. Switching to these electric machines will also lower your energy bills somewhere between a hundred and a thousand dollars—every single year. Those savings make switching to heat pumps and induction stoves affordable. And federal tax incentives and rebates can lower upfront costs significantly.

But clean home appliances won't just save the planet or save you money—they will also save your life. Scientists researching indoor air pollution have made recent discoveries that gas appliances like our stoves are leaking, even when turned off. This gas contains carcinogens, like benzene, which cause cancer. When we burn gas in our homes, it also creates dangerous air pollution, including formaldehyde and nitrous oxide. When scientists measure indoor pollution from something as simple as baking a cake in a gas stove, they are finding pollution levels two times higher than what's considered safe outdoors. We don't have to eat our meals with a side of air pollution. It's time to bring new life to an old catch phrase: "Now we're cooking . . . with magnets and electricity!"

And while this sector is out of sight, out of mind for most people, we can also clean up industrial pollution. Eliminating fossil fuels from all the other parts of our lives will help. With fewer oil and gas operations to supply our energy needs, we cut pollution. And other parts of our industrial system can already be electrified with existing technologies. When we manufacture everything from textiles for clothing to pharmaceuticals for medicine, we can use electricity. That said, creating iron and steel requires high temperatures. With innovation and effort, we can find ways to affordably create heat with electricity rather than fossil fuels. That alone will be a massive breakthrough. When we add up the pollution savings from electrifying both buildings and industry, we can eliminate another quarter of our carbon problem.

With the technologies we have today, clean electricity and electrification can cut around three-quarters of current carbon pollution in the United States. Although the math is not quite the same globally, the basic fact still holds: this is a pathway away from dirty fossil fuels toward clean power.

This transformation will accelerate over time, through virtuous cycles that are both political and technical. Many of these technologies, like solar panels and heat pumps, are democratic—they help everyday people break away from fossil fuels and take greater control over their energy sources. Clean energy and electrification will also create jobs in every corner of the planet. These workers will be vested in the industries of the future, not tied to the dirty fossil fuels of the past. Electricity is called "power" for a reason. It's a tool that can help the people take on the fossil-fuel industry and rival their influence over our politics.

On the technical side, progress also begets more progress. Consider the key clean technologies: solar, wind, batteries, and heat pumps. Over the past several decades, the cost of each of these technologies has fallen rapidly. By the time a statistic is written down, it's already out of date. That's how fast renewables and other clean technologies are becoming more affordable. Scientists call this phenomenon a "learning curve"—the more we produce of something, the better we get at it and the cheaper that technology becomes.

Take for example, solar photovoltaic—the fancy term for a solar panel. In 1957, if we wanted to run an American household on solar power for a month, it would have cost around $300,000. Today, it's just $30. Over the past decade alone, solar costs have fallen by 90 percent. This change is so dramatic it's nearly impossible to chart it alongside other technologies. The line falls so fast it's nearly vertical and seems to slip off the page. This didn't happen because of some invisible hand. It happened because people, in countries as diverse as the United States, Germany, and China, worked together to change laws to make solar cheaper. And people around the world chose to put solar on their roofs. Activism makes markets. Activism drives innovation.

Now the same trend is playing out in lithium-ion batteries—in the last ten years, their cost fell, similarly, by around 90 percent in the United States. Even without government incentives, electric vehicles are now cheaper to buy right off the lot. And the savings just keep coming: EVs cost the equivalent of around just one dollar per gallon to operate. Compared to volatile and expensive oil prices, that is a steal.

Heat pumps are likewise starting to fall in price. And if you're an early adopter, you don't just clean up your own home: you make it cheaper for a neighbor who lives down the block, or even halfway across the globe, to go electric. Heat pumps, like other electric appliances, are also appreciating climate assets: as the electricity mix fueling them gets cleaner, so too do those machines.

And innovation doesn't just hit one point and stop. As problems crop up with new technologies, we will address them. There are already people working to recycle solar panels and lithium batteries. There are folks inventing new battery chemistries. There are people working to make mining more ethical. When we hear stories about the harms posed by clean energy technologies, we should take a beat and ask: Who profits from telling this story? Too often, the fossil-fuel industry is seeding propaganda to make us feel hopeless and defeated. If we delay, they profit. It's easy to forget how bad oil is because it literally disappears into thin air when we burn it. But the lithium in your car is mined once and hopefully used for decades. That makes it more visible, but in the long run, way less harmful. All technologies have impacts. We should not stop asking tough, critical questions. But we also can't forget just how harmful fossil fuels are.

Clean electricity and electrification are the way. It is a pathway from destruction to abundance. You can still have a car, if you want or need one—but you won't fill it up with oil that props up petro-dictators like Putin every time you need to grab some eggs from the store. And when you cook those eggs, you won't be pumping carcinogens into your home through your fossil gas stove. Thinking about the one billion machines we must electrify, in America alone, is overwhelming. But it's also meaningful and motivating. Because with fewer than one billion Americans, every person has something they can do. Ultimately, purpose gives us the highest rewards in life. To know that with our actions we are helping others, those seen and unseen, for generations to come. This is the deepest solace.

The word *electric* has a double meaning. It points the mind toward electricity but also toward thrill and excitement. This march toward the clean electric future is a joyous one, full of possibility. Let's go now together and electrify!

Shared Solutions
Are Our Greatest
Hope and Strength

Gloria Walton

Picture this: Every morning you wake up in an affordable, comfortable home that's powered by sunlight. You walk outside, inhale fresh air, and catch the scent of leaves and trees that keep your close-knit neighborhood cool and safe in the summer. You wave to neighbors picking fresh vegetables in their gardens and smile at kids taking a break from playing in the park to drink clean water from a fountain. You take reliable, public transit to work at a clean-energy job that pays a good wage in a fast-growing industry. You feel pride and joy in what your community has created together.

We all want to live some semblance of a healthy and sustainable life. Yet, too many families wake up in housing they wish they could leave because of neglect, neighborhood blight, underinvestment, or because they are struggling to afford housing in gentrifying areas. Many families step outside their door to be hit with drastic weather and polluted air, wondering where their children can play safely and why there are higher rates of cancer, asthma, and premature death in their neighborhoods. They question why they predominantly have access to jobs that pay so little it seems everybody needs two of them just to cover the bills. The good news is that the solutions and technology to create a regenerative life already exist. What we need to really move us ahead are policies that are grounded in the values of our communities. With love for each other and deep connections to the planet, we can create a future in which every person has a place to call home, and where all people can thrive.

Time and time again, we read reports that reiterate that Black, Indigenous, immigrant, and People of Color communities, and women are hit worst and first by climate change, the side effects of dirty energy, proximity to toxic waste sites, and interlocking systems of oppression. Yet, it is rarely mentioned that these communities at the front lines of the climate crisis are also at the forefront of creating intersectional solutions to tackle an array of issues: protecting their homes and their neighbors; preserving natural ecosystems; building clean and resilient systems for food, housing, energy, and water; and creating local, lasting jobs that pay well and further an economy rooted in care. These communities are showing that a more just and sustainable way of life is practical, affordable, and possible, right now.

The old way of thinking that environmentalism is separate from achieving equity and justice is long gone. People are seeing that the solutions to our problems come from within our communities and ourselves. The most effective solutions connect the dots between our day-to-day lives and the environment. Multiracial and intergenerational coalitions work at these intersections, creating political will to scale solutions that improve our living conditions. People are empowered and coming together to create the change they want to see, the future that reflects their values and visions. And all facets of life—our lives—are made better for it.

Frontline communities across the world are creating solutions that protect natural ecosystems and build affordable, green, and resilient housing. In the US, climate justice solutions tackle the climate crisis and also ensure that we put in place new systems that heal our planet and strengthen our communities:

- In Sunset Park, Brooklyn's oldest Latinx community-based organization, UPROSE, is working with their partners to put solar panels on the roof of the Brooklyn Army Terminal—a large warehouse that occupies ninety-five acres. This means that residents nearby will be able to access cheap and clean renewable energy from the Brooklyn Army Terminal without having to install solar panels on their own roofs, which can be costly and difficult for renters.
- In the predominantly Black community of Britton's Neck, South Carolina, New Alpha Community Development Corporation installed new infrastructure, including hydropanels that will supply water for a community farm and greenhouses and also be used as emergency drinking water when severe weather disrupts the water supply. The installation of these hydropanels will also generate new businesses and create green jobs for the community.
- On the west side of Buffalo, New York, Asian and African immigrants are sharing farming techniques to grow more produce on city lots. They are providing mutual aid during crisis and transforming vacant buildings into affordable, green housing for the elderly.

All these solutions are homegrown, led by members of the communities most impacted by pollution and climate change, demonstrating that people-powered movements work.

Take clean energy, for example: while nearly half of Americans live in a place already committed to 100 percent clean energy, 70 to 80 percent of Black, Indigenous, and Latinx communities live in close proximity to coal-fired power plants and polluting industries. Innovation has brought down the cost of renewable energy, so we know that the technology is here, but deep structural racism in the flow of all forms of capital—public, private, and philanthropic funding—keeps our communities from creating lasting and ubiquitous pathways to the future we want. Adding insult to injury, the innovations born from the conditions we face aren't scaled up, because our communities are too often dismissed by the institutions that uphold power. In the face of the climate crisis, none of us can afford to maintain this status quo, and therefore we must each do our part to shift resources and attention to frontline solutions.

The prospect of letting our natural world provide us an unlimited source of energy by powering all our lives with sunlight and wind shows how humans and nature can coexist. But it's more than powering our homes with clean sources. Clean energy also means improved public health, cleaner air to breathe, cleaner waterways, economic benefits, and more. These issues are as intersectional as our lives are intrinsically connected. We need to fund the solutions created by the most impacted communities at a greater scale, because those solutions are working.

On Navajo and Hopi lands in the Southwest, about fifteen thousand families live in homes without electricity. Native Renewables, an Indigenous-led grassroots organization, is installing solar panels and battery storage in these homes and training Indigenous people for clean-energy jobs in the process. California's Imperial Valley is believed to contain one-third of the world's supply of lithium, an element essential for electric vehicle (EV) batteries. The Lithium Valley Coalition of environmental justice groups, labor organizations, and public health advocates is leading efforts to ensure that lithium extraction does not inflict environmental damage on neighboring communities and ecosystems, and that economic benefits flow to local residents. Instead of following outdated ways of living, we have an opportunity to create a new society, and these communities with lived experiences are giving meaning to intersectional environmentalism—understanding that what affects one of us affects all of us. What solves one problem must not create another.

In his *Letter from Birmingham Jail,* Martin Luther King Jr., writes: "Injustice anywhere is a threat to justice everywhere. We are caught in an inescapable network of mutuality, tied in a single garment of destiny. Whatever affects one directly, affects all indirectly."

The last few years have lit up this truth in neon (LED-solar powered) lights. A global pandemic, economic shocks, global supply chain breakdowns, an escalating climate crisis that's bringing record heat, flooding, and fires—no one on Earth can escape these mutual realities. Our interdependence is clear. Although in the meantime that means shared suffering, it also means that shared solutions can be our greatest hope and our greatest strength.

Capitalistic values have promoted individualistic mindsets and made us believe our resources are finite and competitive. But that doesn't have to be our reality. We have the power to tap into abundance and collaboration. It's our collective responsibility to envision and create the world we want together. We need bold, sustainable solutions that benefit many, not just the few. We can also hold community and grassroots values that nurture a regenerative, healthy, and equitable planet—the values that connect us to our family, our communities, and ultimately to each other.

Community leadership isn't about having it all figured out, it's about the creative spirit, it's about the dream, it's about the solutions created together through collaboration. Not a single sector can solve climate change alone, whether government, media, civil society, or industry. We need to center the values of community, care, and collaboration. This moment requires us to look for leadership in each other and within ourselves. In the wake of natural disasters, we're used to seeing the power of neighbors helping neighbors, keeping one another safe. In a similar way, as threats from the unnatural disaster of climate change intensify, the communities most affected are pulling together to create more sustainable, resilient, and equitable places for people to thrive. It's about solutions created through collaboration with the people around you, the willingness to learn from others' experience and, in return, to share what you've learned.

There's space to dream and create new models of living and relating to each other: inspired by Indigenous wisdom, motivated by people of color's community values, and catalyzed by feminine commitments. It's audacious and it requires tenacity to have a vision for a world we cannot materially see. It takes courage to challenge old ways and build a better future. And it requires love, the source of transformational power—love for ourselves, our people, and the places we call home.

Decolonizing Climate Coloniality

Farhana Sultana

INTRODUCTION

Colonialism haunts the past, present, and future through climate. The burden of climate damage is falling disproportionately on formerly colonized and brutalized, racialized communities in the developing world of the Global South. Frontline communities of the world are feeling climate destruction politically, ecologically, economically, socially, spiritually, and viscerally across the world. This is acutely so in formerly colonized countries across the tropics and subtropics of Asia, Africa, and Latin America that hold less geopolitical and economic power on the global stage. The outcome is a system of climate coloniality where those least responsible for contributing to climate breakdown are impacted more acutely over longer periods. We are still colonized, this time through climate change, capitalist development industry, and globalization, colliding into centuries of varied and overlapping oppressions, yet also concomitant existing and emerging sites of resistance.

The interactions at annual global climate negotiations or COP (for Conference of the Parties) conferences make dramatically evident the climate geopolitics reflecting this imbalance in power and positionalities. At the 2021 COP in Glasgow, colonial tactics were identified and openly called out. While some framed the discussion in terms of climate justice failures, others were more direct in calling out colonial and racial tactics of control and disposal of marginalized communities across the Global South and elsewhere. Many articulated a sense of injustice and climate delay in light of the decades of insufficient critical global action.

The COPs can be seen simultaneously as one of the theaters of climate colonialism (led mainly by corporations, powerful governments, and elites) and also as a site of decolonial, anticolonial, antiracist, and feminist politics (led primarily by activists, youth, Indigenous groups, academics, and unions). While international neocolonial institutions and platforms such as the COPs are resistant to radical change, these are nonetheless also spaces of opportunities to challenge the system, to utter necessary words for more people to hear, collectivize among young and old activists, learn from different positionalities, create new openings and possibilities of alliances—in other words, a repoliticization of climate instead of the depoliticized techno-economist utopias that never deliver. The global theaters of climate negotiations showcase politics and the political, whether subaltern or suburban, where there are both reifications and ruptures in what constitutes politics and its pathways. A sense of despair, grief, rage, suffocation, stagnation, abandonment, and regression coexists with that of revolutionary potentiality, alternative possibilities, collectivizing, determination, worldmaking, and critical hope.

UNDERSTANDING CLIMATE COLONIALITY

Coloniality maintains the matrix of power established during active colonization through contemporary institutional, financial, and geopolitical world orders, and also through knowledge systems. I argue it continues its reach through climate in climate coloniality, which is experienced through continued ecological degradations that are both overt and covert, episodic and creeping—for example, pollution, toxic waste, mining, disasters, desertification, deforestation, land erosion, and more—whereby global capitalism, via development and economic growth ideologies, reproduces various forms of colonial racial harms to entire countries in the Global South and communities of color in the Global North. Thus, climate coloniality occurs where Eurocentric hegemony, neocolonialism, racial capitalism, uneven consumption, and military domination are co-constitutive of climate impacts experienced by variously racialized populations who are disproportionately made vulnerable and disposable. Legacies of imperial violence from active colonial eras live on, not only exacerbating environmental degradation but also increasing climate-induced disasters. As frequencies and strengths of climate-fueled natural hazards such as tropical cyclones grow, the structural violence of colonialism is further experienced and vulnerabilities entrenched. Slow but compounding violence intensifies vulnerabilities that maintain climate coloniality and extend it into the future. Some lives and ecosystems are rendered disposable and sacrificial, fueled by structural forces both historical and contemporary. The racial logic of climate tragedies and cumulative impacts are ever present.

Climate coloniality is perpetuated through controversial global land and water grabs, REDD+ (reducing emissions from deforestation and forest degradation) forestry programs, neoliberal conservation projects, rare-earth mineral mining, deforestation for growth, fossil-fuel warfare, and new green revolutions for agriculture, which benefit a few while dispossessing larger numbers of historically impoverished communities, often elsewhere. Interventions are called by various names and have different tenors—green colonialism, carbon colonialism, fossil capitalism—but often have similar outcomes of domination, displacement, degradation, and impoverishment. Carbon colonialism through carbon-offset projects, which are increasingly ramping up instead of down, despite known critiques and resistances, has been discussed for some time. Extractivism propagated by global capital and state-sanctioned interventions perpetuates geopolitical climate necropolitics within and beyond borders.

As transnational corporate monopolies travel the globe for profit, patterns of colonial dispossession are further entrenched. Extraction and imperialism perpetuate unequal political economies, with imperial and emerging modes of hierarchies of power relations fueled by global market systems.

Climate apartheid is what many call this socio-spatial differentiation in who pays the disproportionate price of climate breakdown, who is made expendable, and who is spared for now. This form of eco-apartheid manifests between and across the Global North and Global South at multiple scales. Climate apartheid exists for those at the intersection of race, gender, and class exposed to ecological harms and toxic environments across sites.

DECOLONIZING CLIMATE COLONIALITY

There is thus an urgent need to decolonize climate to address the harms done and prevent future harm. To decolonize climate at a basic level means to integrate more decolonial, anticolonial, feminist, antiracist, and anticapitalist critiques and struggles into mainstream climate discourses and practices to redress ongoing oppressions and marginalizations. It is not about just recognizing the problems but working toward distributive justice, reparations, and restitution. Decolonizing means accounting for and reflecting on the past and present, in order to configure future pathways to remove colonial and imperial powers in all their forms.

Decolonizing climate would mean rethinking and addressing various institutions and processes at multiple intersecting scales. For instance, it would entail restructuring the world economy to halt the unequal ecological exchange that drains from the Global South to the Global North, which enables the latter's higher consumption and inequitable appropriations. Many climate "solutions" perpetuate the problems of climate coloniality and climate apartheid, so more caution and collaborations are necessary. Likewise, the debates around climate reparations remain contentious, as loss and damage acknowledgment has not been followed through with sufficient financial support.

At the same time, healing colonial and imperial wounds through transformative care, empathy, mutuality, and love holds possibilities. We desperately need to heal colonial wounds everywhere. Ethics of care and collectivity are how we have survived colonialism, capitalism, development, disasters, and disruptions. Caring for each other, despite differences, is what carries us forward through devastations of cyclones, sea surges, riverbank erosion, loss of livelihoods, and degradation of homelands. Nonetheless, it would be callous not to acknowledge the socially mediated, globally and locally produced, ecologically relational vulnerabilities that do worsen over time; how impoverishment and disposability persist; and how increased and repeated harms and shocks make us weary and more vulnerable.

Climate coloniality is thus perpetuated through mundane and institutionalized ways of subalternization of non-Eurocentric, non-masculinist, and noncapitalist understandings of climate, ecology, and human–environment relations. As a result, decolonizing educational systems is fundamental, as systemic cognitive injustices often begin through the formal Eurocentric capitalist education that has gone global. In recent years, the effort to decolonize knowledge and the academy has been powerful in Eurocentric universities. The decolonization of the mind remains critical for epistemic justice and pluriverse, where recuperation of collective memory, dreams, desires, and cultural practices to foster conviviality are important to overcome the colonial matrix of power. Decolonizing knowledge systems to confront climate coloniality requires Indigenization of knowledge and politics. Throughout history, this has been not only ignored, silenced, and resisted in dominant discussions on climate but also often violently oppressed or erased. Yet, power exists in the shadows, forging solidarity and cultural continuity against great odds.

While multiple Indigenous knowledge systems are excluded in hegemonic climate discourses and practices, they are valuable existing cosmologies of decolonial knowledge and resistance that center on accountable, reciprocal, and ethical relations and processes across the globe. There are many different ways that decolonization is enacted, ranging from direct action, law, care networks, leapfrogging alliances, cultural resurgence, and more to center BIPOC futures. For instance, blockades, resistance movements, and landback claims build community claims for liberatory praxis. Speaking in one's native tongue, collective memory and culture rebuilding, retelling of historiographies, and celebrating human–nonhuman kinship are some of the strategies. Native singing and dancing are resistance, and valuing storytelling is decolonial action. Reclaiming sacredness is anticolonial, and counter-stories and counter-mapping are strategies of opposition. Defending territorial ontologies is decolonial politics. Recognizing relational entanglements and healing fosters well-being and convivialities. For many, various practices are simultaneously coping mechanisms, refusals, resistance movements, and decolonial actions, where recollections of collective memories and practices as well as enactments for liberation remain the goal.

Through such processes, ethics of care, care networks, and prioritizing collective well-being instead of only individual well-being become more clarified. This accounts for embodied, ecological, economic, and political safety from harm and fosters flourishing. Healing the colonial wound through transgressive love and solidarity becomes possible. Alienation is fought against by reclaiming sacredness and relationalities, by moving toward liberation and self-determination without apolitically fetishizing or romancing the local communities or cultures.

SOLIDARITIES AND POLITICAL LIBERATIONS

What is evident is that political liberation from climate coloniality will rely on allyship and solidarities in intentional anti-imperial and anticolonial projects across peoples of occupied, postcolonial, and settler-colonial contexts—particularly among BIPOC from across continents. Political consciousness informed by anticolonial politics is necessary for decolonization and abolition of systems of harms. The natures of these relationships need to be worked out, but coalitions come together by working through contentions and differences. Kinship building can be fraught; it needs humility and humanity, overcoming alienation, and acknowledging differences and commonalities to build shared goals.

Decolonization thus must build political community and practical solidarities that foster pluriversality and reparative relations, ethics of care, and restoration of humanity and agency in the battle against climate change and climate coloniality. The ruthless extractions and dispossessions across territories everywhere showcase the connections across place-based materialities to broader extractive ideologies and colonial-capitalist greed. Indigenous scholarship demonstrates the importance of self-determination and ecological kinship, more-than-human relationality, and multispecies justice. Recognizing and valuing living complex ecosystems and agroecology, instead of marketized nature as commodity in a capitalist exploitative system, become vital for epistemic and material climate justice.

Ultimately, there is no single blueprint for decolonizing climate, as decolonizing is a process and not an event; it is ongoing unlearning to relearn. It is the many acts, small and large, acting in constellations and collectivities over time and place, that bear results.

An Indigenous Systems Approach to the Climate Crisis

Jade Begay

A colleague recently told me that climate justice is about build-ing ties between people, their land, and their traditional, ancestral ways. In all my years of doing environmental work, this is one of the most succinct ways I've heard to describe what climate justice means for Indigenous people and communities: reconnecting to our land is an integral piece of addressing climate change for both our Nations and our wider communities.

To understand why, we need to take a closer look at how the last few generations of Indigenous peoples have been removed from our lands and lifeways, and how far-reaching the conse-quences of those actions have been. When settlers arrived and colonization began, as NDN Collective program officer PennElys Droz explains, our economic systems were targeted for disruption and destruction:

Removing a peoples' means of providing for themselves is a cunning way to suppress and control them. George Washington famously led the burning of Haudenosaunee seed houses. The United States encouraged the slaughter of buffalo to destroy the ability of the Plains Nations to provide for themselves. And in California, settlers methodically destroyed the oak trees that the people depended upon. A state of dependency was intentionally created, with the Nations having to look to their colonizers for survival assistance.

When Indigenous peoples across North America were forced into boarding schools, they were subjected to assimilation and punished for speaking their languages, and their cultures were vilified. At the same time, the Indian Removal Act gave president Andrew Jackson the authority to negotiate with Tribes in the South to relocate west of the Mississippi, so that white settlers could develop those lands. When the 1956 Indian Relocation Act was signed into law—part of the Indian Termination era—Native people were encouraged to move to urban areas with promises of better jobs and prosperity, though, more often than not, they were met with the oppressive cycles of systemic poverty and discrimination.

The policies enacted in the nineteenth and twentieth centuries didn't just violate Indigenous rights and sovereignty, dissolve treaties, and eliminate reservations. They also made way for the extraction of resources by oil and fossil-fuel corporations from our Native and Tribal lands. But even though connections between the climate crisis and the removal of Indigenous peoples from our lands are stark, they are often overlooked. If we are to save our planet and its resources, we must be willing to interrogate this harmful, pervasive connection and work to repair it at the source.

We know now that the mismanagement of forests by several government agencies, including the US Forest Service, in a misguided attempt to protect trees, has led to worsening wildfires across Western states like California, Montana, and New Mexico. And because capitalism—powered by the fossil-fuel industry—prioritizes profit over human life and ecological well-being, we see this story play out repeatedly across the country, whether it be pushing pipelines through sacred lands and water systems or destroying old-growth forests in Alaska, which are some of our best carbon sequesters.

Removing Indigenous peoples from our land took away our ability to carry and pass on traditional ecological knowledge, such as how to manage lands, our connection to traditional food ways, and our traditional economic structures. There is much to learn, for example, from the Indigenous peoples who have practiced controlled and deliberate burns that restore ecosystem-wide health. Recently, the Yurok people in California have partnered with local fire departments to bring back the ancient practice of controlled burns, which allowed hazel to grow in the area for the first time in many years. The success of this partnership demonstrates the importance of centering Indigenous people and our knowledge of the planet in the fight against the climate crisis.

"Every Indigenous Nation had a traditional economy," PennElys Droz explains:

> A way of gathering and distributing what we needed to live and thrive, that was connected to extensive trade routes across the Americas and allowed for the exchange of the gifts of the land, knowledge, language, and culture. These economies developed based on countless generations of learning from our homelands and each other, learning to care for the beings that give us life while

ensuring their continuance. These economies also reflect-
ed an understanding that our homelands are living beings to be engaged with in good relationships, in order to
receive the blessings of abundance, and the importance of
keeping good relations and resource distribution among
community members.

Whether it was two hundred years ago or today, when we
exploit, extract, and/or pollute Indigenous lands, we destroy the
critical knowledge and technology that is needed to manage the
climate crisis.

BUILDING REGENERATIVE INDIGENOUS ECONOMIES

NDN Collective is a systems organization that uses Indigenous
frameworks and knowledge to tackle issues like climate change. We
know that when even one part of a system is not working, the entire
system is impacted. Therefore, we work holistically and with solutions front and center: as we advocate and mobilize our communities to end their dependence on the fossil-fuel economy, we are also
resourcing and investing in them to create pathways to reconnect
with their lands, culture, and traditional knowledge. We take this
approach because we know that the results will foster climate justice and allow Indigenous communities to take back the power that
comes with having a mutually beneficial relationship with our land.

Despite generations of being attacked and forced from our
lands, Indigenous-led organizations and brilliant individuals are
helming innovative efforts to fight back, furthering climate justice
by connecting Indigenous people to our rightful land and ancestral
knowledge. In Cordova, Alaska, for example, Native Conservancy
is working to build a regenerative Indigenous economy as it battles
mining that could devastate both the traditional lands of the Eyak

people and the regional ecosystem. A case in point: in 1989, Dune Lankard watched as the *Exxon Valdez* tanker hit Bligh Reef and spewed tens of millions of gallons of crude oil into Dune's homelands and ancestral waters in Alaska's Prince William Sound, an event he refers to as the "day the water died." Since the devastating oil spill, Dune has worked enthusiastically and tirelessly as a community leader, protecting more than a million acres within the Cooper River Delta and 1,500 miles across the Pacific Ocean to Kodiak, an area permanently protected from development. And by building a regenerative economy, Dune and the Native Conservancy are working to strengthen the inherent rights of sovereignty and subsistence and the climate resiliency of their community. At this moment, Dune's team is hard at work building a community kelp seed nursery and conducting research for kelp cultivation. The goal is to build the first Native community–run kelp restoration project in the *Exxon Valdez* spill zone, and, by supporting Native ownership of wild kelp seed, to empower Native villages to manage their own means of growing kelp.

Kelp provides more than a source of food—as Dune puts it, "What you get is a product that can be used for a variety of purposes such as nutraceuticals, compost, fertilizer, biofuels, and many other byproducts":

> In addition to being a food and excellent fertilizer for on-land farming, kelp offers immense benefits to the environment. It improves water quality, provides valuable habitat for hundreds of ocean species, and it has the potential to mitigate climate change impacts such as ocean acidification. Kelp can also serve as an incredibly effective carbon sink; estimates suggest that kelp forests can sequester five times the carbon dioxide of terrestrial forests. In other

words, growing kelp is a win-win-win: for our Native communities, our ailing blue planet, and for creating a new resilient, restorative, regenerative economy.

INDIGENOUS PEOPLE TRANSITION TO RENEWABLE ENERGY

Operating in the rural Navajo Nation, Native Renewables—founded by Wahleah Johns, who recently began a new role as the senior advisor at the Office of Indian Energy Policy and Programs—empowers Native communities by giving them access to renewable solar energy. Not only does Native Renewables address the challenge and infrastructure gap of Native and Tribal communities lacking access to electricity by installing off-grid solar PV systems, they also build capacity in our communities by leading workforce training. Since its founding, Native Renewables has cohosted several classroom and hands-on workshops in Arizona, New Mexico, and Alaska on solar energy technology and how this form of renewable energy can support our communities.

Native Renewables integrates cultural values and teachings into their work by bringing in Indigenous cosmologies about the sun and providing tools that connect solar energy education and language revitalization. Navajo Nation can and will continue to be a model for the rest of the nation, as they continue transitioning from fossil fuels. Native Renewables is supporting this transition by equipping their communities with the skill sets and resources to achieve true energy sovereignty and liberation from the fossil-fuel industry.

CENTERING INDIGENOUS COMMUNITY

At their root, the Green New Deal and other regenerative solutions and progressive policies around climate justice are actually Indigenous models. As we continue to mobilize for comprehensive climate justice legislation, and as we begin to see more recognition of Indigenous traditional ecological knowledge in federal policy making, it is imperative that we center Indigenous communities and grassroots leaders who have been on the ground implementing place-based solutions for decades. Indigenous people know what it takes to save our planet and the life-giving resources it provides.

How the Ants Moved
the Elephants in Paris

Renato Redentor Constantino

To live with hope in a world that seems determined to race off a cliff: this is the real radical choice. As the late John Berger counseled, it is to live with history where the past is a dear companion instead of a mere instrument used to bludgeon adversaries.

Nowadays, the 2015 Paris Agreement is the most familiar example of global solutions to the climate crisis. This global treaty was twenty-one years in the making, with many powerful governments and industries working to dilute what was necessary. It was birthed by an insurrection that grew from the ashes of the Copenhagen negotiations in 2009, which failed to secure a meaningful agreement. In the six years after the unsuccessful Denmark climate conference, the powerful exerted tremendous effort to keep a tiny number, 1.5, out of United Nations documents. One and a half degrees Celsius represents what science advises as the maximum allowable rise in average global temperature relative to preindustrial temperature levels. Beyond that lurk menacing outcomes. For a huge number of developing countries, in particular the coalition of island states, the group of Least Developed Countries, and the Climate Vulnerable Forum (CVF), fractions of a degree mattered. The figure of 1.5°C versus 2°C meant the difference between survival and annihilation.

Many considered 1.5° unachievable not because of science but of politics. Representatives of elite international nongovernmental organizations (INGOs) viewed the 1.5° limit with disdain. Some even lobbied leaders supporting vulnerable country governments to quash efforts to advance the more ambitious climate goal. They claimed that this was in order not to harm "global" consensus. On the other side, a few leaders of social movements that stridently supported the 1.5° limit in principle sneered at the notion that developed countries would ever be persuaded to adopt the target. Intending to secure global accord based on acceptably low common denominators, they judged calls to limit temperature rise to 1.5° as impractical, divisive, and harmful. But vulnerable countries, in particular government members of the CVF, saw things differently. They considered 1.5°C—the ability of hundreds of millions to survive and thrive—as the only credible basis for unity.

The CVF, a cooperation association founded in 2009, composed of governments of countries acutely vulnerable to climate change, commissioned three reports, which found that setting a target of 2°C was tantamount to legitimizing human rights violations on a long-term global scale, preventing the full enjoyment of the right to food, health, shelter, and water. In response, throughout 2015, campaigns by groups such as the Vulnerable Twenty (V20) intensified and called for a reset of ambition.

Against the position taken by rich countries and many INGO leaders, the CVF rolled out a global campaign that highlighted gains should warming be limited to 1.5°C. For instance, half of all coral reefs worldwide may yet survive at 1.5°C, whereas warming that breached 2°C could see all of them lost. The CVF campaign showed, as well, how warming to 1.5°C or 2°C would require the same technologies and transition programs except that to reach the lower target, action would need to be undertaken earlier and more rapidly.

For vulnerable countries, 1.5°C meant full decarbonization of the global economy and that no one would be exempt from action. Everyone, including countries such as the Philippines, Bhutan, Ghana, Barbados, and Guatemala, would need to deliver their fair share of climate action while big emitting countries would need to do far more and act faster.

CVF's global campaign convened senior ministers from Africa, Asia, Latin America, the Caribbean, and the Pacific. It anticipated spirited resistance from leaders who continued to dismiss calls for a 1.5°C-anchored outcome in 2015 by harnessing popular power from grassroots groups supporting the more ambitious warming limit and by reaching out to trade unions, youth groups, celebrities, scientists, grassroots coalitions in developing countries, and civil society networks.

Instead of wilting under the pressure of powerful voices calling for moderation and low ambition unity, CVF urged countries "to heed the 2013–2015 Review" insisting that "immediate measures to strengthen the goal to the below 1.5 degrees Celsius target are indispensable to the . . . survival of a number of our nations, and the prosperity of our populations, people everywhere, and the world." CVF called "on all nations to seize . . . [the] opportunity [in Paris] for a climate-secure framework that keeps warming below 1.5°C."

Unfortunately, findings from the 2013–2015 review never made it to the Paris plenary. On its last session on Friday, December 4, Saudi Arabia succeeded in marshalling influential developing-country allies to block submission of the review's findings for plenary debate, aided by the silence of developed-country governments.

But CVF was prepared for such an outcome.

Over a hundred governments had by then expressed official support of 1.5°C, and it was CVF's turn to divide and conquer. Germany and Canada had broken ranks to publicly stand for 1.5°C, and they would soon be joined by more and more rich nations while more and more bystander governments, none of whom wanted to be left out, signed up their support.

As the review came to its close, the soft-spoken senior Filipina physicist Dr. Rosa Perez, IPCC member and Philippine delegate to the global warming treaty negotiations, gently read into the record her statement:

> Our discussion has taken a bad turn. There is no question for the vulnerable countries who are losing 2.5 percent of GDP and fifty thousand lives each year that we condone an outcome that commits us to undermining our survival. We have to really question at this point the basis for not forwarding the draft text to the COP given the scale of the work undertaken and the gravity of the decision. The parties who stand in the way of recommending a sound decision based on the information available will be remembered by the children of today for the failure in Paris and we will shout it from the rooftops.

Soon, the Eiffel Tower would light up with a message blazing, "1.5°C." By the middle of the second and final week of the negotiations, even Saudi Arabia, on behalf of the six Gulf Cooperation Council countries, would cave. The Saudis would request a parley with CVF to ensure they would not be singled out as the global villain in the endgame.

Even the Vatican, whose delegation to the Paris talks had steadfastly carried the 2°C line, was not spared. I was in the room in a small closed-door bilateral on December 8 with Philippine officials, led by secretary Emmanuel de Guzman, delegation head of the Philippines, which chaired CVF from 2015–2016, along with the Vatican's delegation to the negotiations.

As the meeting drew to a close, Monsignor Bernardito Auza, Apostolic Nuncio to the United Nations, finally agreed to support 1.5°C, saying, "Free the text. Take down the brackets," in the negotiating text that was preventing the inclusion of 1.5°C in the final outcome.[5] De Guzman had requested representatives of the Holy See to support the CVF's position. After they had visibly hesitated, I said, "Your Excellencies, God will show the way."

The High Ambition Coalition for Nature and People (HAC) has often been highlighted as an alliance that ultimately helped cement full support for the Paris Agreement, which is true.[6] What few know is that, without CVF, the HAC would not have been born. It was a privilege to have worked closely with Tony de Brum, who represented the Republic of the Marshall Islands, one of the atoll nations most at risk from rising sea levels. De Brum was a giant of a man, not only because of his considerable height and heft but also because of his colossal spirit and sharp mind.

De Brum and I had early exchanges late in the first week of Paris in December, when frustration ran high and the specter of failed talks was spreading concern among negotiators. De Guzman had accepted De Brum's invitation to help form a multicountry coalition of the willing, the first meeting of which, December 7, I was asked to join on behalf of the Philippine's De Guzman. Toward the end I took the floor and said to other delegates that the Philip-

---

5 CVF press release, "Holy See joins final push in Paris for 1.5°C," December 10, 2015.

6 HAC is an intergovernmental group of more than one hundred countries, co-chaired by Costa Rica and France and with the United Kingdom as Ocean co-chair, championing a global deal for nature and people with the central goal of protecting at least 30 percent of the world's land and ocean by 2030.

pines appreciated the invitation and would welcome joining more meetings if invited, but that we, much less the CVF, should not be considered members of the coalition. Later, De Brum and I would huddle and agree that the meeting's composition was a problem— the only developing countries represented were the Marshall Islands and the Philippines, and the rest were countries from North America and Europe. "This is not a proper coalition," said De Brum. I nodded and replied that it would remain so until vulnerable-country priorities were adopted. De Brum nodded back.

The following day the Philippines was invited again, and once more I had to repeat for the record that neither the Philippines nor the CVF should be considered members, adding that far more might join the coalition-to-be if only it would endorse, in writing, 1.5°C.

On Wednesday, December 9, the meeting became heated. Earlier that day, to my surprise, conference monitors throughout the sprawl of the UN climate camp in Paris displayed a list of the HAC members, including the Philippines. I quietly shared with De Brum what I intended to say just before the HAC meeting began. De Brum smiled and replied, "Go for it."

Ten more developing-country representatives had joined the room, but most of the people were still from developed nations. Around fifteen minutes after the meeting had opened, I asked for the floor and said,

> This is not easy to say but it needs to be said. We've always been clear we remain observers but also that we'd be happy to join if the vulnerable country agenda is supported. Earlier today, the Philippines was erroneously listed on monitors as a member of the HAC. This group is thus faced with two choices. It can now support 1.5°C in writing,

knowing when it does so the Philippines will actively reach out and encourage CVF members to join the HAC, or it can maintain the status quo and assess our commitment to publicly denounce this coalition.

In the room were the lead negotiators of Canada, members of the EU, and the US. After about ten seconds of silence, a red-faced Miguel Cañete, representing the EU, replied tersely to say that my statement was noted. Those few words were tantamount to surrender. Sometimes the ant's best defense against an elephant is firm resolve and clever strategy.

I remember meeting De Brum in the corridors the next day. He looked me in the eyes and said softly, "They gave in." De Brum gave me the most powerful bear hug I've ever encountered in my life, and we parted briefly to hold each other's arms and look at one another, then we laughed aloud once more and embraced and laughed again.

Much of the winter of 2015 remains a blur to me, a stack of images and sounds shuffling randomly as if someone was rapidly flipping the pages of a big book. But two instances have always remained vivid: the bear hug of Tony de Brum and my last night at the Paris talks.

I remember eagerly heading for the exit of the UN grounds in the Le Bourget conference center. It was late, and the cold was bracing. Exhausted and thinking of my bed near Gare du Nord, I piled on the overcoat, buttoned up, and wrapped a scarf around my neck. I walked briskly through corridors linking the warren of provisional COP21 structures that had housed tens of thousands of debates. Just before the last doorway, something gave me pause. I stopped and looked up slowly to my right, squinting at the massive sign framing the German Pavilion: "Below 2°C—Together We'll

Make It!" I thought, "Well, they never expected our insurgency."

The gavel signaling the official conclusion of the Paris talks needed one more day to thresh out a last set of details, but that night the outcome of the climate negotiations was already clear. A tipping point had been breached, and the common cause that people had championed would land where it had been aimed.

By December 12, what for years was considered a romantic but ultimately doomed threshold limiting warming below 1.5°C had become enshrined as the benchmark of global ambition. Not in a footnote, not as an impotent rhetorical "whereas," but in Article 2 of the Paris Agreement.

We need stories to remind us why hope is complicated but necessary, because the opposite mode is to live neat lives powered by a self-affirming wireless fidelity to all-terrain gloom, where all signs point to defeat and despair waits at every turn. To hope is to embrace uncertainty, knowing the bad guys have not won yet.

On the first days of the Paris talks, I drafted an opinion piece with the brilliant leader of the CVF secretariat, Matthew McKinnon, with Monica Araya—the dazzling virtuoso behind the Costa Rican group Nivela—and with fierce communicator Cindy Baxter and eminent scientist Bill Hare, both with Climate Analytics. Our words ring true today.

We declared, "No one in the world believes for a minute any of this will be easy, and that it will not be without resistance or setbacks, but everybody knows that the obstacles to action are essentially purely political. This now is the central challenge of Paris—for political ambition to rise to the levels required for a major breakthrough in climate action."

The work is far from finished. Many industrial nations are still approving fossil-fuel projects as if keeping warming to 1.5°C didn't matter. As always, it is up to us—citizens of the world—to do what is necessary, despite, or in response to, the bleak prospects we are facing today. As I wrote in the introduction to the international literary anthology on global warming, *Harvest Moon: Poems and Stories from the Edge of the Climate Crisis,*

> Shadows haunt us for what they conceal as much as what they suggest . . . an elemental fact [that reminds us] when the sense of hope feels eclipsed . . . it is time to cast our own penumbral light . . . Across the night sky the waxing moon grows. *Luna crescens,* which describes less the moon's heraldic shape and more its stage in keeping with the original sense of crescent in Latin. To come forth. To swell. To thrive and grow. Because this is our moment. We are the crescendo, able to exert our own gravitational pull. And we have stories to share.

To Hell with Drowning

Julian Aguon

I know nothing of the night sky.

This saddens but does not surprise Larry Raigetal, a master navigator who is chewing betel nut beneath a canopy of stars. He is from Lamotrek, an outer island of Yap, in the Federated States of Micronesia. But we are meeting in a canoe house on the neighboring island of Guam, where I call home. As we speak, Raigetal is using his hands to split the horizon into a thirty-two-point star compass. He is drawing on centuries of knowledge to explain to me the art of wayfinding—a method of noninstrument navigation that has been used by his people for thousands of years to voyage between the many atolls and islands of Micronesia.

To my surprise, the compass he is conceptually grafting onto the sky is more than a map of stars as they rise and fall from east to west across the horizon. Wayfinding is a manner of organizing an elaborate body of directional information collected and committed to memory by countless navigators before him and passed down through chants to his grandfather, to his father, to him. It's a living repository of spectacularly specific details about sea swells, wind currents, reefs, shoals, and other seamarks—

including living ones. A pod of pilot whales. A shark with special markings. A seabird.

As a Pacific Islander, I knew that the canoe house has long been a place of learning, and I'd come to ask Raigetal whether wayfinding had been compromised by climate change. As a human-rights lawyer working at the intersection of Indigenous rights and environmental justice, I'd also come because I believe that the peoples of the Pacific have important intellectual contributions to make to the global climate-justice movement. We have insights born not only of living in close harmony with the Earth but also of having survived so much already—the ravages of extractive industry, the experiments of nuclear powers. We have information vital to the project of recovering the planet's life-support systems.

Finally, I'd come because my personal and professional reserves were depleted. Like so many others working in the climate space, I'd been feeling overwhelmed since August, when the IPCC released part of its sixth assessment report. The conclusions were bleak. Reading the report felt like being buried alive by an avalanche of facts—the facts of sea-level rise and progressively severe storms, among others—and I was looking to claw my way out.

As the darkness deepened around me and Raigetal, I realized two things. First, the climate-justice movement must listen more carefully to those most vulnerable to the ravages of climate change, such as Oceania's frontline communities. Second, we who are waist-deep in that movement need more than facts to win. We need stories. And not just stories about the stakes, which we know are high, but stories about the places we call home. Stories about our own small corners of the Earth as we know them. As we love them.

In my corner, Micronesia, the facts are frightening. We are seeing a rate of sea-level rise two to three times the global average. Some scientists theorize that most of our low-lying coral-atoll nations may become uninhabitable as early as 2030. Faced with the prospect of climate-induced relocation, some leaders have contemplated buying land in other countries in anticipation of having to move some or all of their people.

One leader has already sealed a deal. In 2014, the then president of Kiribati, Anote Tong, entered into a purchase agreement with the Anglican Church for more than five thousand acres in Fiji, paying nearly $9 million for them. (Kiribati has since begun using the land for farming.) Though the deal was seen as visionary by some, to others it marked a kind of death. After all, at what point does an agreement that envisions the relocation of an entire human population—now some 121,000 people—become more eulogy than contract?

In Fiji, the government keeps its own kind of death list—an official record of all the villages that may have to be relocated because of sea-level rise. Using internal climate-vulnerability assessments, the Fijian government has determined which of its coastal villages are most susceptible to coastal erosion, flooding, and saltwater intrusion. As of 2017, forty-two villages were on the list. If and when they are forced to move, they won't be the first: in 2014, Vunidogoloa formally relocated to higher ground, some two kilometers inland.

When I spoke with Sailosi Ramatu, that village's headman, in July, he told me the move was hardest on the elders. In the months leading up to the relocation, they held prayer circles. They fasted. They readied themselves for the rupture of having to abandon their ancestral lands. In Fiji (as in many of our islands), the people are tethered to the earth, as enshrined in the concept of *vanua*, a word that means "the land" and "the people" at once. Vunidogoloans live and love and die on their lands, most of which they do not even own, at least not as individuals. Rather, theirs is a system of communal land ownership. They tend to their gardens. They bury their dead. They even bury their umbilical cords. So, it was no surprise when, as about thirty families set out for the new site, some of the older women wailed as they walked.

Perhaps that's a sound the sea makes when it rises: old women wailing.

Not everyone made the journey; the dead remain interred in a cemetery at the old site. According to Ramatu, one of the biggest struggles his people faced was leaving their buried loved ones behind. Some worry they'll be cursed for abandoning their deceased relatives. Others walk around with holes in their heart. Like the old man who visits the cemetery nearly every day to sit by the grave of his dead wife. I'm not sure which flowers he brings her, if any. But I imagine they're beautiful.

Perhaps the story of climate change is a story of flowers.

These are the facts in the Republic of the Marshall Islands (RMI), the country where the US military houses its Ronald Reagan Ballistic Missile Defense Test Site. There, a crucial study on sea-level rise found that coral-atoll nations may not be able to sustain a human population past the present decade. This conclusion was met with trepidation by the Marshallese people I spoke with, who hear the ticking of the climate clock louder than most.

The 2018 study, led by the United States Geological Survey and commissioned by the Pentagon, focused exclusively on an island in the Kwajalein atoll that supports some 1,250 American military personnel, contractors, and civilians living there and on nearby islands. For the most part, the US otherwise ignores this region. Wake Island, where an additional study on sea-level rise is now being done, is proof of that fact.

Wake, an island with no permanent inhabitants that the US considers an unincorporated territory, is run by the Air Force under authority of a caretaker permit issued by the Interior Department.

For its part, the RMI not only has a competing vision of what care-taking looks like; it also has a competing claim to Wake. In April 2016, the RMI formally claimed Wake Island when it filed its maritime coordinates with the United Nations secretary-general.

The truth is that neither government is entirely correct. The strongest claim is that of the Marshallese people themselves, who say the island is theirs by way of history, culture, and birthright, and who long to be able to take proper care of it. They also say that Wake is not the island's true name.

Its true name is Enen-Kio. The island of the orange flower.

Famous in lore for the beauty of these flowers, Enen-Kio is also known for its rare assemblage of nesting seabirds—frigates and albatross, among others. Legend has it that local warriors, seeking to prove their worthiness, would journey to the island in search of the wing bones of one such seabird. Fourteen years ago, on another starry night, a high chief explained to me that the retrieved bones were used as chisels in traditional tattoo ceremonies.

I did not grasp the significance of the strip of orange splayed across the RMI flag until much later. Former president Hilda Heine would tell her poet daughter, Kathy, who would tell me: for the Marshallese, orange is the color of bravery.

On my island, climate change is a story of storms. Guam—the largest and southernmost of the Mariana Islands and an unincorporated territory of the US—lies within one of the most active regions for tropical cyclones in the world. The typhoons that have historically battered the island are so strong, they're often called "super typhoons."

Everyone here remembers their first. Mine was Omar, in August 1992. We were unprepared—my mother, brother, sister, and me. This was in part because my father, who typically did the preparatory work of putting up shutters and removing debris from around the house, had recently died. I remember the four of us huddled behind a cream-colored mattress. I remember tracing its embroidered flowers with my finger.

I remember everything, really. Trees and telephone poles cracked in half. The roof of our neighbor's house went flying, as did his canopy and one of his cars. I remember glass everywhere, as several windows and a sliding door shattered. I remember the sound of the wind as it blew under the bottom of my bedroom door. Like an old man sucking his teeth.

Pamela is the one my mom remembers. May 1976. One of the most intense storms to strike Guam last century, Pamela generated eight-meter waves and ravaged the beaches on both the northern and eastern sides of the island. She sank ten ships in the local harbor. She did an estimated $500 million worth of damage. But none of this is what my mom remembers. What she remembers, what she will never forget, is a single white toilet. American Standard. The one thing left of her house when Pamela was over.

Then there was Paka. December 1997. The wind and rain beat down on us for twelve hours. The barometric pressure was so low that it was believed to have induced labor in nine pregnant women. Paka, like Russ in 1990 and Yuri in 1991, unearthed untold numbers of dead bodies when it slammed into the southern cemeteries of Yona and Inarajan. Corpses spilled out of their coffins. Coffins bobbed like buoys in the bay.

Several families spent weeks combing the beaches in search of their loved ones. Some were never found. My aunt, who worked for one of the cemeteries, said that one family was able to identify their father's body only because of a cherished baseball cap, which they had buried him in and which had stuck to his skull by way of a mess of seaweed. Suffice it to say, when the IPCC dropped its latest report, confirming that tropical cyclones are just going to get stronger, my corner of the world shuddered.

After all, although 1.5°C of warming will make these storms even more severe, that same severity will increase dramatically with 2°C, let alone 3°C. I can't begin to imagine what any of this will mean on the ground—and not just for Guam or the Northern Mariana Islands, but for Vanuatu, Fiji, and Solomon Islands, whose communities already seem to be lurching from one Category 5 cyclone to another.

Throughout Oceania, the story of climate change is also a story of ingenuity. In the Carteret Islands—off the coast of Bougainville, in Papua New Guinea—the women are taking matters into their own hands. Frustrated by how slowly the Papua New Guinean government was implementing its relocation plans, they sought to mobilize the community around the issue of relocation. They formed an organization and named it Tulele Peisa, which means "sailing the waves on our own" in the local Halia language.

To date, Tulele Peisa has organized several community consultations as well as visiting missions between the Carteret Islanders and potential host communities in nearby Bougainville. According to Ursula Rakova, the group's leader, Tulele Peisa has also secured several tracts of arable land on which it is now growing gardens of taro and cassava. She told me the group has planted more than thirty thousand cocoa trees and even established a cocoa-bean refinery. All told, Tulele Peisa has developed an eighteen-point relocation plan for its community. Rakova said other coastal communities have followed suit and are currently formulating their own relocation plans.

So, it would seem that the Pacific Climate Warriors were right. The youth-led group fighting climate change across Oceania, as part of the global 350.org network, famously declared: *We're not drowning. We're fighting.*

And we are.

The Marshall Islands has spearheaded the Climate Vulnerable Forum, a group of forty-eight countries that works to amplify voices that have long been marginalized in the climate realm. Fiji, too, has taken a leadership role, presiding over the Conference of the Parties for the Twenty-Third United Nations Framework Convention on Climate Change and spearheading the so-called Talanoa Dialogues—sessions that use storytelling to foster more empathetic decision-making.

One could argue that Fiji is also leading the way on the complex issue of climate-induced relocation. Consider the list of forty-two villages slated for possible relocation. However heartrending, the very existence of the list is a testament to Fiji's efforts. Conversations about relocation are enormously difficult to have, but that country is having them.

Tuvalu, Tokelau, Kiribati, and the Marshall Islands have joined forces with the Maldives to form a coalition of coral-atoll nations to advocate for the financial resources necessary to adapt to climate change. To date, what money most of them have been able to secure has been limited to funding the first-generation stuff of seawalls and early-warning systems—nowhere near the level they will need to actually adapt, let alone adapt in place. But they press on, planting mangroves and plugging away at their national plans.

A group of law students from the University of the South Pacific led a different charge in Vanuatu: they advocated that that climate-vulnerable country take the lead in pursuing an advisory opinion on climate change from the International Court of Justice. As these students see it, the lack of clarity around state duties is impairing the collective efforts of the international community to respond effectively to the climate crisis. In September, the students succeeded in their initial goal, and Vanuatu announced that it would spearhead the initiative. (I should note here that I am leading the global team assisting Vanuatu in this effort.)

In the Marianas, many Indigenous Chamorros and Carolinians of Guam and the Commonwealth of the Northern Mariana Islands are fighting the destruction of our lands and seas by the single largest institutional producer of greenhouse gases in the world: the US military. On land, these activists are opposing the construction of a massive firing range, which they say will destroy a limestone forest and imperil a whole host of nonhuman life. At sea, they're challenging the Defense Department's attempt to militarize a section of the ocean almost the size of India.

All this to say, if my corner of the Earth had an anthem, it'd be this: To hell with drowning.

That anthem was never more clearly on display than during the twelfth Festival of Pacific Arts, held in the summer of 2016, when the *Lucky Star* arrived at the local harbor. *Lucky Star* was one of three canoes sailed to Guam from Lamotrek by the wayfinder Raigetal and his crew of apprentices. What made this particular canoe so special was the fact that its sail was traditional, meaning it was woven from pandanus leaves by the women of Lamotrek.

There, in the weeks leading up to the voyage, it was discovered that the knowledge of how to weave such a sail was very nearly lost. Only one woman still knew how to do it.

Her name was Maria Labushoilam, she was a ninety-year-old master weaver, and she was dying. Maria would spend the last two weeks of her life teaching fifteen women how to make that sail. From her deathbed, she taught them how to harvest, dry, and split the leaves, and then how to weave them. After she died, the women completed the sail without her. The community raised it together.

The Lamotrekese are emblematic of the predicament all Oceanic peoples are facing today: we come from cultural traditions rich in beauty and resilience—the same traditions that have enabled us to thrive in our ancestral spaces for thousands of years—but that is simply not enough to ensure our continued survival. The part simply cannot save the whole. The answer to the question of climate change must come from everyone, or it will come from no one.

You could say Maria performed something of a miracle in her final days. With nothing more than pandanus leaves and love, she opened a window to a world—a future in which good people refuse to simply lie down and die, a future rooted in respect for possibility, a future with room for us all.

May we have the courage to climb through it.

An Extremely Incomplete List of Climate Victories

1974
- Chipko movement to protect trees in Uttarakhand, India, stops the destruction of the forest near Reni village after a four-day standoff.

1985
- In Brazil, the National Council of Rubber Tappers forms to protect both workers and the Amazon rainforest; in the US, Rainforest Action Network is founded to protect rainforests globally.

1987
- The Montreal Protocol, a global treaty to stop using ozone-depleting chemicals is achieved, potentially preventing more than 443 million cases of skin cancer and a full degree Celsius of warming.

1994
- The United Nations Framework Convention on Climate

Change enters into force, thus cementing a process of nations coming together regularly to negotiate a global response to climate change.

1999

- Activists around the world protest the World Trade Organization meeting and shut it down in Seattle, succeeding in stalling its assault on workers' rights, environmental protections, small farmers, and poor nations, setting the stage for a global climate movement.

2002

- Middle school teacher turned California legislator Fran Pavley proposes California vehicle emissions standards that, in 2009, become national standards.

2004

- Kenyan social, environmental, feminist, and political activist Wangari Maathai wins the Nobel Peace Prize, the first time climate/environmental protection is recognized as peace work. Her Greenbelt Alliance has planted more than 50 million trees.

2005

- The first global day of climate action is held in response to the Kyoto Protocol going into force. The Kyoto Protocol commits nations around the world to reduce greenhouse gas emissions.

- Friends of the Earth drafts a resolution for greenhouse gas emission reduction for the UK that passes in 2008 as the Climate Change Act.

2006

- Former vice president Al Gore's climate-science film *An Inconvenient Truth* reaches broad audiences.

2007

- Launch of the Great Green Wall project in Africa, which aims to grow 8,000 km² of vegetation across the Sahel to tackle desertification.

2008

- Ecuador adopts a constitution enshrining the Rights of Nature, upheld in a 2021 suit against mining companies.

- Ørsted, previously known as DONG Energy, begins its transformation from a coal-based energy company to a wind-based company, launching offshore wind farms that by 2020 include the world's largest, producing 1,218 megawatts of power.

2009

- First meeting of the Climate Vulnerable Forum, an international partnership of nations most at risk.

- October 24 marks the first International Day of Climate Action, with more than 5,400 actions in 181 countries, including underwater in the Maldives and atop a mountain in Antarctica.

2010

- In British Columbia, Canada, the Wet'suwet'en hereditary chiefs and their supporters set up a camp directly in the path of the Enbridge Northern Gateway Pipelines, which will culminate in huge 2020 protests and continues today.

- In Germany, renewable energy generates more than one hundred TWh (billion kilowatt-hours) of electricity, providing nearly 17 percent of national supply.

2011

- Campaigns against the proposed Keystone XL Pipeline begin to gather force through building alliances, public awareness, actions, and more.

- Ecuadoran judge finds Chevron liable for $8.5 billion in damages to Indigenous inhabitants of the Amazon.

2012

- Fossil-fuel divestment movement begins and spreads globally—by 2022, more than $40 trillion will be divested.

- First Nations Canadian women launch the Indigenous movement Idle No More to protect nature and culture.

- German solar power grows by more than 7.6 gigawatts (GW), breaking the previous records of 7.5 GW in 2011 and 7.4 GW in 2010.

2014

- Pacific Climate Warriors use canoes and other traditional vessels to blockade the world's largest coal port in Australia.

- Four hundred thousand people march in New York City to press for stronger action at the upcoming global climate talks in Paris. More than two thousand additional actions occur around the world.

- New York State bans fracking (made permanent in 2021).

2015

- The Paris Climate Treaty, a legally binding international treaty, is adopted. It sets the goal of limiting global warming to 1.5°C or below, largely because of organizing from vulnerable countries.

- Indigenous resistance succeeds in canceling the Pierre River tar sands mine in northern Alberta, Canada.

2016

- The Standing Rock Sioux Tribe sues the US Army Corps of Engineers to stop the Dakota Access Pipeline and wins a temporary halt late that year. A great gathering arises to protest the pipeline and protect the water.

- Thanks to Indigenous organizing, Canada's Northern Gateway tar sands pipeline is canceled.

- In another Indigenous victory, the Lummi Tribe defeats the proposed Cherry Point coal terminal in Washington State.

- New York State denies water permit to Constitution Pipeline Company for a proposed natural gas pipeline, and activists succeed in returning the seized land to landowners.

2017

- April 21: Britain's first ever working day without coal power since electrification.

- In a US milestone, the number of coal mines has plummeted from 1,435 in 2008 to 671.

- First Nations groups force cancellation of TransCanada's Energy East tar sands pipeline, which would have been North America's largest, carrying more than 1.1 billion barrels of tar sands daily.

- Māori Tribe wins their case to gain legal personhood for the Whanganui River, which the group recognizes as an ancestor.

- In Turkey, grassroots resistance forces cancellations of two coal plants in the Aliağa region.

- The Stop Adani Alliance is launched, representing 1.5 million Australians taking a stand against Adani's proposed mega coal mine. Through years of organizing the project is delayed and halted, and, especially thanks to the leadership of the Wangan and Jagalingou traditional owners, more than 4 billion metric tons of carbon pollution have been avoided.

2018

- August: Greta Thunberg, age fifteen, begins FridaysForFuture in Sweden, leading to youth climate actions on six continents.

- Sunrise Movement launches in US, demands a Green New Deal.

- Extinction Rebellion launches in UK.

- After decades of resistance, activists in Brittany, France, defeat plans for a major airport.

- Ireland votes to divest public funds from fossil fuels.

- New York State activists defeat Constitution Pipeline Company's methane gas pipeline.

2019

- Responding to the calls of youth climate strikers, a record number 7.6 million people take to the streets for the largest day of climate action ever.

- Waorani people of Pastaza win a historic ruling in Ecuadorian court, protecting half a million acres of their territory in the Amazon rainforest.

- After much organizing in Kenya, a major coal plant near the historic town of Lamu is halted.

- 2.25 gigawatt Navajo Generating Station shuts down; at full power it burned fifteen tons of coal each minute, twenty-four hours a day, making it one of the United States' largest greenhouse gas producers.

- Green New Deal introduced in Congress.

- More than fifty US coal companies have gone bankrupt in the decade ending in 2019.

- The UK becomes the first major economy in the world to pass laws to bring all their emissions to net zero by 2050.

- A Polish district court condemns project to build a 1 gigawatt coal-fueled power plant.

- A Dutch court orders the country's government to reduce carbon emissions by 25 percent. The nonprofit Urgenda Foundation had first filed their case in 2013.

2020

- Standing Rock Sioux Nation wins lawsuit against Dakota Access Pipeline.

- UK goes sixty-seven days without using coal to generate energy.

- Texas gets more power from wind than coal for the first time, and Iowa nears 60 percent electricity generation from wind.

- Headline: "Indigenous-Led Resistance to Enbridge's Line 3 Pipeline Threatens Big Oil's Last Stand."

- Residents of North Carolina, Virginia, and West Virginia succeed, after a six-year campaign, in canceling the Atlantic Coast Pipeline.

2021

- European Investment Bank announces it will end all loans to oil and gas firms.

- International Energy Agency says (at last), "There is no need for investment in new fossil-fuel supply in our net zero pathway."

- On a single day, May 26, a Dutch court orders Shell to cut emissions by 45 percent; 61 percent of Chevron shareholders vote to reduce the oil company's emissions; and climate-positive Exxon shareholders oust two company directors.

- At COP26, thirty-nine states, including Canada, Ethiopia, Fiji, Finland, France, Gabon, Iceland, Italy, Mali, and New Zealand, and financial institutions sign a treaty to end support for fossil fuels flowing abroad by the end of

2022 and instead prioritize finance for clean energy.

- Keystone XL Pipeline is finally killed on President Biden's first day in office. The US also rejoins the Paris Agreement.

- Stop Cambo campaign, started to fight the Cambo oil field, expands to take on all new oil and gas fields in the UK.

- A Polish district court invalidates plans to build a 1 gigawatt coal-fueled power plant, thanks to shareholder activism.

- A new report demonstrates that "Indigenous resistance has stopped or delayed greenhouse gas pollution equivalent to at least one-quarter of annual US and Canadian emissions," equivalent to 1.8 billion metric tons of CO_2.

- Jordan Cove LNG (liquefied natural gas) export terminal and 229-mile Pacific Connector gas pipeline in southern Oregon are canceled after opposition by Indigenous peoples in the region, including the Karuk, Yurok, Klamath, and Round Valley Tribes.

- The government of Greenland announces it will stop all oil and gas exploration because it "takes the climate crisis seriously."

2022

- Concentration of ozone-depleting substances in the mid-latitude stratosphere falls more than 50 percent, back to levels observed in 1980, before ozone depletion was significant, thanks to the 1987 Montreal Protocol.

- Ecuador's Supreme Court gives Indigenous nations the right of final approval over extraction projects affecting their lands.

- Colombia elects climate champions president Gustavo Petro and vice president Francia Márquez, pledging to put "the defense of life above the interests of economic capital."

- African court rules in favor of Indigenous land titles and reparations from the Kenyan government.

- Brazil's High Court is the first to declare the Paris Agreement a human rights treaty, and Brazil reelects Luiz Inácio "Lula" da Silva, a huge victory for protecting the Amazon.

- Australia finally votes out its coal-backed conservatives in favor of a more climate-positive administration.

- One spring day, California gets almost 100 percent of its energy from renewables. In summer the state boosts its energy goals to 90 percent renewable/clean by 2030 and 95 percent by 2035.

- US judge cancels oil and gas leases of more than 80 million acres in the Gulf of Mexico, because climate change was not sufficiently considered when permits were issued.

- Austria and Germany initiate super-cheap transit passes and tickets to save fuel and prevent private-car use.

- Court in South Africa strikes down Shell's offshore permit for oil and gas exploration.

- Hornsea Two project becomes world's largest offshore wind farm, overtaking Hornsea One nearby.

- Philippines' national commission on human rights declares climate change a threat to human rights and invokes the culpability of governments and fossil-fuel corporations in violating those rights.

- The US Congress passes its first major piece of climate legislation (with both good and bad provisions), under the title Inflation Reduction Act.

FRAMEWORKS OF POSSIBILITY

When you fall apart, don't forget to love the pieces.

—**Bayo Akomolafe**

May we let the sadness come and teach us how to live. Let it be the mud for the lotus, as Thay [Thich Nhat Hanh] says. Let us sit with it and let it pass through us so that it might be transformed to something like love.

—**Ocean Vuong**

If we let ourselves feel this, we will be better for it. We will wake up and reach out and finally tap into our love for one another and our planet.

—**Reverend angel Kyodo williams**

We must offer ourselves—and each other—space to grasp onto that rest, joy, possibility, and freedom now, or we'll grind ourselves completely away simply surviving the oppressive pressures around us.

—**Rachel Cargle**

What to Do
When the World Is Ending

Yotam Marom

I am part of a generation that feels, constantly, and even in the most mundane moments, that the world is ending. Almost every article I read these days begins with the same preamble listing all of the overlapping crises, topped off by the climate crisis, which will quite possibly lead to the extinction of our species. We have been told so many times that we have an extremely brief window to turn things around, and even then, it is already too late in many ways. I feel swallowed by despair, and I know I'm not alone.

But while there are some things about this moment that feel unique, I remind myself that *the experience of the world ending is not new.* Whether due to a prophecy or a very real looming threat, many of our ancestors also likely felt that the world was ending. And in many cases their worlds *did* end. The devastation on Easter Island, the fall of Carthage, the arrival of Columbus, the centuries of chattel slavery, the destruction of Hiroshima, the cold war, even the Cuban missile crisis—these all must have felt like the end of the world. Facing loss, despair, uncertainty, and death is as much a part of the human experience as anything else.

It's true that this notion of historic solidarity might not be encouraging. But perhaps it is useful in another way, can point us toward some wisdom we have yet to unlock. Maybe it can shed some light on a question it would serve us well to answer: *What do people do when their worlds are ending?*

During the Holocaust, which my own family narrowly survived, Jews were often herded into ghettos across Europe to be starved, brutalized, murdered, and ultimately transferred to concentration camps, where they were either gassed or worked to death. In the Warsaw Ghetto, there was a famed underground youth-led Jewish resistance that rose up and resisted the Germans for twenty-seven days, against all odds. On their final day, in a holdout encircled by German soldiers, some of the last remaining fighters jumped out of windows to their deaths.

Letters that the ghetto fighters wrote to one another document that the fighters didn't hold any illusions about their position. They knew they would not win. In his final letter, Mordechai Anielewicz, the twenty-four-year-old commander of the Warsaw Ghetto Uprising, writes: "It is impossible to describe the conditions under which the Jews of the ghetto are now living. Only a few will be able to hold out. The remainder will die sooner or later. Their fate is decided." The fighters were mostly starving teenagers with limited weapons and almost no training, facing an almost endless line of German soldiers with tanks and machine guns. They knew that the ghetto would be burned to the ground. Yet, Jews facing the end of their world did unimaginably brave and kind and selfless things. And not just the ones who took up arms, but the many along the way, who simply did what so many people feel called to do when they are confronted with suffering. The ones who carried a letter from one place to another, hid a refugee, fed a person on the brink of starvation, took in a lost child, snuck an extra ration home for a sick elder, gave up their blanket for someone colder.

But why? Why did they take these risks if they knew they would lose? Why bother if the world was going to end anyway? There is this part of Anielewicz's letter as well: "The fact that we are remembered beyond the ghetto walls encourages us in our struggle." Many ghetto fighters hoped that their actions would inspire others to rise up as well, and they did. In fact, they inspired me, decades later and half a world away. Undoubtedly, these teenagers, who only a few years earlier went to school and took scouting trips and played ball in the courtyard, had become real strategists. They thought about how to organize people, how to help people feel the belonging necessary to build strong groups, how to connect to a bigger purpose than what was in front of them, how to fight. They understood, it seems, that purpose, strategy, and action could overcome despair.

Yet, there seems to be more here than strategy—something about self-worth, purpose, *meaning*. Viktor Frankl, a psychotherapist, author of *Man's Search for Meaning*, draws on his own experience in a concentration camp. Frankl writes: "There is nothing in the world, I venture to say, that would so effectively help one to survive even the worst conditions as the knowledge that there is a meaning in one's life." Those most likely to survive the camps, Frankl tells us, were those who had a reason to keep fighting to live—a loved one, God, socialism, a vision of a future world they were fighting for. They saw a reason to keep going, so they took agency, even if only in tiny ways—like searching hard for that extra calorie to make it through one more day. Frankl writes: "Everything can be taken from a man but one thing: the last of the human freedoms—to choose one's attitude in any given set of circumstances, to choose one's own way."

Despair is a reasonable reaction to the world we live in. I feel it every day: as I watch my five-year-old daughter fall in love with parts of this world that will disappear in her lifetime; when she gives her pocket change to an old man on the street who is cold and hungry in a world that will almost certainly become colder and hungrier before it becomes anything else; as I read the neverending onslaught of news articles on everything from rising white nationalism to melting glaciers. It takes enormous effort to break through it, and even then it comes back. Despair is the kind of thing that comes in waves, creeps under your skin, finds its way into your belly when you're not paying attention.

But I do break through it from time to time. Often it's after having really engaged with the heartbreak of it all. Sometimes it's when I feel glimmers of hope from small victories we've won along the way. But most often it happens when I am pushed or pulled by others who have decided to keep fighting—young people dropping out of college to fight for the Green New Deal, Indigenous elders blockading pipelines in the freezing cold, Black people organizing their communities against police repression, and more. They see the writing on the wall as well as I do, know the science just the same—know it in numbers but in the losses of their communities, too. *They know their worlds might end,* and yet they choose to take action. As George Lakey, who has spent his whole life fighting, and whom I have prodded about hope before, reminds me one day on the phone, in one of my own moments of despair: "I can let the newspapers tell me how my life will go, or I can decide for myself."

In the moments of wisdom encouraged by these heroes, I remember that despair is my vanity talking. It is an indulgence in the illusion that what is here and now is inevitable, that the future is written, that we can see how it will unfold. Despair is not about reality, or the world, or even ultimately the people we care about. *It is about us.* It is the act of allowing our very real sadness and fear to limit our sense of what is possible, about finding safety and comfort in that darkness, about avoiding heartbreak. Despair is the easy way out.

Despair is also, quite simply, bad politics. By surrendering the fight outward, despair refocuses us inward. It encourages what I've called the politics of powerlessness, marked by navel-gazing, endless process, posturing, and the internal power struggles and call-outs that weaken our organizations and movements. When we don't believe we can win, we reach instead for the comfort of being surrounded by people who think and talk and look like us, the thrill of being part of the in-group, the small pleasures of being right and pure. In despair there is no need for good strategy, no need for healthy group culture. These are things we only need if we intend to take a real shot at winning. Despair is a self-fulfilling prophecy; it blocks us from taking agency, which makes it all the more likely that our worst fears will come to pass.

As we look out on the political landscape before us, we have every right to assess it as bleak. But nothing about it is inevitable, and we shouldn't expect our tiny human brains to know how everything will unfold. There are undoubtedly major social upheavals before us. The deep crises we are in the midst of will bring not only pain and suffering but also incredible opportunities for change. People will find themselves moved, outraged, seeking, and out in the streets again in great numbers, many times in the coming decades.

Rather than pretend we know how it all ends, we should do the things we know have worked before: nurture and join powerful social movements, and build institutions that provide masses of people with a vehicle for belonging, meaning, and long-term struggle. This requires good strategy, healthy groups able to wield it, and a movement with a culture open and creative and compelling enough to win over the enormous numbers of people necessary for real transformation. And it requires humility—about what's possible, about ourselves, about each other. "Wherever human beings are, we at least have a chance," James Baldwin reminds us, "because we're not only disasters; we're also miracles."

But beyond strategy, there is also just the simple, humble, profound task of *being authentically alive on this planet in a time of collapse.* Here, too, there is action, because there is more life in the taking of agency than in watching it flutter past us. Taking agency makes us smile and laugh and cry. It gives us the chance to express love and rage. It pumps our blood and fires our synapses. It creates new possibilities, compels action in others, and creates connection, which is what movements are made of. It gives us the opportunity to practice incredible traits like heroism, generosity, and care, lets us experience the joy, love, and gratitude that go hand in hand with those traits.

As Arundhati Roy writes: "There is beauty yet in this brutal, damaged world of ours. Hidden, fierce, immense. Beauty that is uniquely ours and beauty that we have received with grace from others, enhanced, reinvented and made our own. We have to seek it out, nurture it, love it." Yes, there is beauty to uncover still. It won't disappear the despair, or grief, or heartbreak, but it can, perhaps, prevent us from sinking into them.

So, what do we do when the world is ending? The same things that so many of the giants on whose shoulders we stand did when *their* worlds were ending. We choose to face our despair—to walk toward it and through it—choose to take action, choose to build movements. We do it because we don't know how it ends, because there are possibilities out there that we simply can't see from here. We do it because every person organized and campaign won and fraction of a degree of global warming prevented will save lives. Because movements that believe are far more powerful than movements that don't. And, yes, we fight because fighting is one of the ways we get to nurture our courage and generosity and hope and all those other fundamentally human traits that we treasure most—because our lives will be infinitely richer in that struggle than outside of it. We do it because it is how we get to truly live.

Meeting the More and the Marrow

WHAT MORAL ANGUISH, GRIEF, AND FEAR GIVE US

Roshi Joan Halifax

Some time ago, I read these words from climate activist and storyteller Terry Tempest Williams and thought that this is one brave and upright woman. Williams wrote: "A good friend said to me, 'You are married to sorrow.'" And Williams replied, "I'm not married to sorrow, I just choose not to look away."

When it comes to our climate catastrophe (and all else, for that matter), I feel that Williams is so right: we mustn't look away. We are in a crisis of the mind and a crisis of the heart. And the nature of this crisis goes to the very core of how we live as social beings, how we live as moral and caring beings, and how we meet fear, how we meet loss, and how we meet grief. And as well, in how we open to hope, and how we open to wonder. But before we touch briefly on the givens of hope and wonder, I think that we must navigate, a least a little bit, the tough geographies of fear, and of grief, and, as well, of moral suffering, to discover what these harsher landscapes might offer us.

To begin, I feel it is so important for us to realize that our climate catastrophe is a source of moral anguish for many who are aware that the disastrous shifts in our climate are related to the extractive views and institutions that feed structural and direct violence toward our Earth and toward Earth-cherishing peoples and people of color. We also are aware that our climate catastrophe has a deep impact on the rights and well-being of all lives on our Earth and is a situation that directly affects our character and our experience of integrity and safety.

There is more: if there is to be a viable, morally grounded, and healthy future, timely and committed social and environmental action is essential, as is psychospiritual transformation. In spite of the pessimism generated by our conventional media, we see that in actuality many individuals have chosen to cultivate the qualities of heart and mind that make it possible to see the destruction we have wrought on this Earth and have the guts to not only bear witness but also are socially and environmentally engaged. We are also seeing that many have nurtured the courage to recognize the grief and shame associated with the climate catastrophe and are taking action to transform the habits of mind that bind one to consumerism, racism, and elitism, and blind one to the harms that have been caused.

In the midst of all this, I believe that it is essential that we grieve. We have to come to terms with the multiple losses that we are experiencing at this time and to move out of the parched landscape of denial to experience grief and learn what grief can give us. I also believe that it is important at this time for us to work with our fears, those sharp and foreboding edges that keep us knotted in the fist of threat.

We need to grieve and face our fears not only individually but also collectively, as communities of care and purpose. Yet, our society struggles with grief and fear, and often regards both as signs of weakness, as something to be ashamed of, as something to be denied, or something to be hidden away, or something that should be processed as quickly as possible.

This has been complicated by the fact that many of us have lost the myths, the stories, the rituals, and the individual and collective practices that support us in transforming these difficult and rich experiences. I also believe that our inability to face and work with grief and fear has other implications. For example, this inability can contribute to social fragmentation, increase social and psychological polarization, and give rise to fundamentalism and even violence.

Grief and fear are human responses to loss—loss of social connection, loss of autonomy, and loss of certainty. And now we face the loss of the stable ecosystems in which human beings have lived during the nearly twelve thousand years of the Holocene. And fear? Fear is a response to a real or imagined threat. And at this time, we are certainly in a situation of threat, if not worse. So, grief and fear are our frequent companions.

Yet, I think it's also important to know that like grief, fear can be a kind of doorway. My old friend Joseph Goldstein wrote, "When fear arises, it means that we're at some edge of what we're willing to be with, what we are willing to accept. Right there is precisely the most interesting place of practice because that is where we have set a limitation, a boundary for ourselves. If we can see that and recognize it, then that is the place to work, to look, to explore. That is the place to open."

When we find ourselves in the grip of fear or in the depths of grief, we can experience these states as a powerful invitation to understand that we are at a threshold that we need to explore. We also might realize that, if we avoid experiencing fear and grief, more suffering might well happen. Each time we turn away from our grief, we might harden our anguish. Each time that we turn away from fear, we can increase the grip of dread. And, in doing so, we may miss the opportunity to strengthen and resource ourselves with what we learn from these painful landscapes, so that later we can face with humility and hard-earned wisdom what is happening in our changing world.

Returning to the words of Terry Tempest Williams: "There's deep beauty in not averting our gaze. No matter how hard it is, no matter how heartbreaking it can be, it is about presence, it is about bearing witness." Williams added, "I used to think bearing witness was a passive act. I don't believe that anymore. I think that when we are present—when we bear witness, when we do not divert our gaze—something is revealed—the very marrow of life. We change. A transformation occurs. Our consciousness shifts."

So perhaps we can discover that fear and grief are givens. Working our grief, facing our fears, can transform us. Hard as it may be to see this when we are in the thick of the experience of grief or in the grip of fear, both can be deeply humanizing, both can deepen our empathy, and both can increase our capacity for compassion and insight.

In working honestly with grief and fear, we can bring presence into whatever is before us. In this regard, I want to mention something that is important to keep in mind. Even though the institutions we serve in, the situations we encounter in our everyday lives, the people we are connected to, or our very Earth might seem to be devolving at this time, from the point of view of complex adaptive systems, even as our world and our lives can appear to be so fraught, systems that break down have the potential to reorganize themselves at a higher, more robust level of functionality when we learn from the breakdown.

This process applies to people as well. The experience of breakdown, of being swamped by grief, fear, or moral anguish can give one a very deep and positive view of the potential of others and ourselves to grow from ordeals instead of being diminished or crushed. This refers to the benefit we can derive from the psychological changes from our struggle with challenging life circumstances. These experiences can foster extraordinary resilience; they can foster hope; they can also foster awe and joy. People who have survived trauma can come back transformed by the experience and see that suffering has made them hardier, rather than more fragile, with the ability to thrive in the present rather than being overwhelmed by the past.

Transforming our suffering doesn't mean that we are going to be returned to the state that we experienced before. But we can discover that suffering and loss have given us a greater ability to live in the present rather than to be constantly overwhelmed by the past. It is also helpful to remember that what we have been through can make it possible for us to be more connected with presence, and can nourish wisdom and greater compassion as well as humility. Beyond the ending of the old way of being, there is space for the emergence of the new, and to imagine a future in which the wounds are still there but in a form that makes one wiser and humbler and supports one to flourish.

The novelist Haruki Murakami wrote in his *Kafka on the Shore*, "And once the storm is over you won't remember how you made it through, how you managed to survive. You won't even be sure, in fact, whether the storm is really over. But one thing is certain. When you come out of the storm you won't be the same person who walked in. That's what this storm's all about."

This is the experience of transformation that Terry Tempest Williams wrote about. From this perspective, thriving means that we experience growth beyond survival. Through what we have allowed ourselves to go through, the very marrow of life is revealed.

Through this, we experience a deeper and braver connection to all of life, the hard and the beautiful, the tough and the mysterious, and we meet all of what life gives us, like when the Buddha met Mara, the embodiment of evil: "Hello, old friend, I know you."

Some of us receive the precious opportunity in this time to use the struggles that we are experiencing to dedicating ourselves to fostering sanity, care, and justice in the world. We have heeded the call to abandon futility and meet our moral anguish, our grief, and our fear with openness and curiosity. We have also allowed ourselves to be worked by the power of adversity in order to meet the unfolding and uncertain present with inquiry, hope, awe, and loving action.

And if we can't, then we do not turn away from that. Sometimes we have to pause, not ready to take the next step. Sometimes we make unfortunate mistakes and withdraw from the world in shame. Sometimes we falter in the midst. Sometimes we fall apart and stay that way a long time. And sometimes we need to step away, to retreat, to take the backward step. It is simply not our time to step forward. But know that we, too, are being worked. And others are being worked in their own way. It is not to add the weight of judgment onto the burden that we are already carrying. It is not to turn away from our current experience, even if our response does not meet our so-called standards. It is rather to meet it with, "Hello, old friend. I know you."

And for those of us who encounter moral anguish, grief, and fear head on, may we also come to meet our tangled world and realize that this is sacred work, as astrophysicist Adam Frank suggests. He writes: "It goes back to the very roots of being human, to a time when our hunter-gatherer ancestors could feel the sense of *more* when they came on a bend in a river or stumbled across a mountain glade. The sacred is the opposite of all those times when we are living in our heads, mulling over our worries, or focused on just trying to get by. Sacredness appears in those moments when life overflows its banks—when we see the vast, variegated, and infinite network of life and being we each are part of." I believe that this *more* is another given when we do not divert our gaze, when we do not turn away from anguish, when we do not turn away from this world, this Earth, from each other, from ourselves. This *more* of the sacred rests in the very marrow of our lives and as well rests between us and all beings and things.

Bigger Than
the Easiest Answer

Interview with Kathy Jetñil-Kijiner

Thelma: I feel like you wear many hats: you're a poet, climate envoy, leader, mother. You are also an artist who recognizes the intersectionality of so many issues. Not only do you write about climate change, you also talk about nuclear history and the Marshall Islands, racism, and so much else. For you, what role does art play in our movements for justice?

Kathy: I think art is there to inspire us and to reflect back the more complicated feelings about these issues. I think sometimes in these spaces it can be so easy to just say it's either this or that, it's either one or the other. This is the solution. So, we have to go with it, and you know we're not allowed to have emotions about those solutions that are sometimes better for the overall movement. But art allows us to kind of take that space back, step back and be like, "Can I just grieve for a second? Can I just feel sad and sorry that this is happening and that we're here, and that we have to make these decisions?" I think that's what it does for me. It gives me that sort of space to be meditative, and to be bigger than the easiest answer.

Thelma: And with your lens as a mother—one of our pivotal poems that you shared at the UN in 2014 was "Dear Matafele Peinem"—a poem to your then newborn daughter. Now she's grown up and is a tall kid. Looking back, what has it meant to be a mom in this age of the climate crisis?

Kathy: I guess the biggest thing is learning to unplug. There are moms who are social justice moms, who definitely teach their kids about the issues. And I do too. I've laid out the bare facts as much as I can on different issues for her, so that she has my perspective on it. And I always back it up with "This is my perspective. You're going to learn about it a bit more probably on your own, and you're welcome to come up with your own solutions, or come up with your own opinion, and we can talk about it more."

But I also feel like mothering has taught me that I need to just tap out—stop and not be doing that work—and to be completely present with her, and just have fun with her. Have fun with her in this world together, and that that relationship is the most important relationship above and beyond anything else that I'm doing. My work defines me and it's a part of who I am, and I love it. I love the work that I do. But I also know that there's no one to replace her. She's so fun and she's so cool, and I don't want to ever lose that relationship with her.

Also being a mother meant I had to grow in a lot of ways. I had to let go of a lot of patterns of thinking, and I had to do a lot of inner work that has nothing to do, absolutely nothing to do with climate change. I had to do a lot of inner work to be more patient, to be more kind, to heal a lot of childhood trauma that I had so that I don't pass it on to her, and that took a ton of work. I'm still working on it. Being a mom forces you to grow.

And as far as "Dear Matafele Peinem"—that's my most hopeful poem, because you're writing to a baby, and of course you want to be hopeful. So, it's always been my most hopeful poem every time I perform it. I do enjoy performing it, even if it's a really old poem. I've done it a million times. I enjoy performing it because it becomes a refrain. It's like a prayer. It's like me saying and willing it to happen for her. So that's the kind of magic that I see behind that piece. I'd like to believe that saying it over and over the way that I have been will make it real.

Thelma: I love what you said about just being present. I think, as climate activists we can be so focused on the future that we forget the joyous moments of the present.

Kathy: I think that's definitely a problem. When we put the work before our kids, then it just becomes—well, what's the point of all this if you're not going to enjoy your kid, if you're not going to enjoy this time with them. But you've got to make sure not to lose sight of that focus. And that's a lot easier said than done.

Thelma: You said "Dear Matafele Peinem" is your most hopeful poem. Being from the Marshall Islands, one of the low-lying atoll nations on the front lines of the climate crisis, at real risk of rising sea levels, what is your relationship with hope, and how has it shifted over the years?

Kathy: I think Julian [Aguon] was someone who really made me kind of interrogate the idea of hope—when it becomes weaponized, in a sense. It struck a chord with me—the way hope can be weaponized, the way it can be used to not face the reality of the situation, to not be able to grieve, to not give time for grief, to not give time for those of us who are really feeling it first.

There is a sanitization of the movement at times, because people can say, "This isn't about just your island. This is about everyone who is vulnerable, not just the Marshall Islands. You know we have to stay hopeful. We don't want to be too dark, too depressing." Those are things I've heard before.

That's why I love art—it can be as depressing as we want. I guess, in terms of hope, it really helps when I know that it's not just me. I remember I had a journalist ask me, "What's it feel like to have this thing on your shoulders?" They asked me that at the COP [Conference of the Parties annual global climate summit]—and it was so easy. I was like, "It's not on all my shoulders. It's not just me—there's a whole team." There's a whole domestic team back home implementing and working through all the kinks at the national policy. There's consultants who come in and who genuinely care about the issue, and have important insights that they can share with us. There's another envoy and having ministers, too, that support us. Hope, for me, is recognizing that it's not on one person, that the savior complex is completely self-destructive. There was a little bit of a way in which I was raised to have that kind of savior complex. "You're going to get your degree, and you're going to go home and help your people." There was a lot of pressure in that, and I completely disagree with it now. I think it's really self-destructive to think of yourself as a one-man band. For a while it did feel like that. I felt burned out really easily.

Hope for me is being able to work within our team. Just one step at a time working on it daily and knowing that we're doing something. We're working toward something, knowing we have these strategies we're coming up with. Being active like that really helps me personally with not losing hope, with not getting swallowed by the enormity of the task.

It is really hard to keep hope when you look at our island. I've seen the way people look at us after I've given them the background of our island, and what we're facing. It's like they look at you kind of with pity—they're like, "Oh, you're fucked." They think about it privately, they might even crack a little joke about it to try to be edgy, because they're uncomfortable. And, you know, I don't think we are. I think we're fucked in that we're already going to have to make changes. And we shouldn't have had to make changes at all. I think people always take that grief for granted, that part of the grief, that we've already made up our minds and this is about to happen. There's totally grief there, and I feel like everyone always passes over that realization really quickly. But, we're not going anywhere. I think there's something really exciting about that. We've all talked as a team, and we're "absolutely not." We're not leaving, we're not giving up. Like-minded folks who are older than me, also younger than me, and we're all like—"No, absolutely not." This is a nonnegotiable. That's really exciting for me. That's probably what gives me hope, is that I'm not alone in thinking that.

Hope for me has to be practical, and it has to be grounded. As a poet, there should maybe be something more magical about my hope. But I have found that it's easier for me to bear it and not lose my focus if I keep my hope grounded and practical. I'm just going to solve this next task, which contributes to this overall plan, which will hopefully keep us in place.

The Asteroid and the Fern

Jacquelyn Gill

In September 2018, I found myself in the vast expanse of the Siberian taiga. It was my first time north of the Arctic Circle, and I was traveling through time. We were chasing rumors of incredibly well-preserved ice age specimens—not the tiny pollen grains or dung fungus spores I usually work with in my forensic reconstructions of vanished landscapes. These were reported to be entire organisms, complete with skin and flesh and fur, perfectly preserved for tens of thousands of years by the permafrost. I'd spent more than a decade researching ice age ecosystems, but I had never actually seen one except in my mind's eye, a mental tableau populated by long-dead species and a vanished climate. I had flown nearly around the globe to get as close as I could to the past, hoping that the mammoth-filled Pleistocene could reveal clues about our own warming world.

As I descended into the glittering darkness of a permafrost tunnel, I felt like Orpheus searching not for a lost lover but for a past that had always remained just outside my grasp. And I soon found it: frosted teeth and bones of mammoths, woolly rhinos, steppe bison, and horses, jutting from the frozen ripples of dark soil like chunks in a freezer-burned ice cream. The surface was dotted with clumps of ancient grasses, still green—and even entire trees in some places. I was walking through a nearly intact ecosystem, frozen in time. Some inhabitants, like the cave lion and the mammoth, had vanished forever. Others, like the caribou and the lark, had survived. What was the secret to their resilience? Were they just lucky spinners on fortune's wheel? Or did they know something I didn't about surviving an apocalypse?

I've held countless mammoth bones, teeth, and tusks. I've dedicated the better part of my career as a scientist to understanding the ecological consequences of climate change and extinction. But it wasn't until my journey into the Siberian permafrost that the past became real to me. These weren't specimens in a laboratory; they were individuals. They had hearts that beat like mine and lungs that expanded to resist death, even at the last. Among the remains of those doomed species gathered for millennia in the long dark, I mourned, but I also marveled at the incredible resilience of the survivors. And, as my breath frosted in the light of my headlamp, I promised to honor the dead by serving the living: the muskox and the rhinoceros and the larch trees who were their kin—and ours.

When I emerged from the permafrost tunnel, I knew I was not the same person who had entered. The proximity to past upheavals—so much death and data concentrated in one place—had changed me. And like Orpheus, I could not resist the urge to look back—not because of the ghosts I'd left behind in that glittering underworld, but because of what waited to greet me outside. It's the present that has always called me back to the past; it's the living who compel me to speak with the dead. As I climbed up and out, I stepped into the light of the same Arctic sun that had set on the last woolly mammoth some 3,700 years ago, even as it rose on my own ancestors, half a world away.

. As a paleoecologist, my work takes me to some of the worst moments in Earth's history, from apocalyptic asteroids to the climatic disruptions at the end of the last ice age. On a bad day, the fossil record reads like a casualty list five billion years long. We even divide the geologic timeline by the branches that drop off the tree of life, marking the moments when a particular group of organisms disappears from the fossil record forever. Indeed, the planet has hurtled from one catastrophe to another like a giant pinball, careening from icehouse to hothouse and back again: oceans starved for oxygen, great volcanic arcs spewing greenhouse gases, the deadly cloak of an impact winter, mass death.

Yet every time we roll toward oblivion, the flipper sends us back into the game, and the scoreboard dings and flashes as biodiversity ticks upward once again. In the aftermath of every catastrophe, the survivors radiate into a diversity of forms and strategies, weaving new patterns in the tapestry of life. It is a story as awe inspiring as it is sobering. The fossil record reveals just how far we can push an ecosystem before it collapses like a Jenga tower, or how suddenly an ice sheet or an ocean current can turn on a super-heated dime. The rocks compel us to remember, not only that we do not repeat the past but also that we understand what it took to build a present worth protecting.

The very worst mass extinction in Earth's history took place at the end of the Permian, some 251.9 million years ago. It was so utterly world shattering that scientists call it the Great Dying: roughly 90 percent of life on Earth was lost—a literal decimation. Even the insects, which tend to emerge from mass extinctions relatively unscathed, were hit hard. The end-Permian extinction was caused by an enormous injection of atmospheric greenhouse gases by a group of volcanoes known as the Siberian Traps (ironically, not far from where I had my permafrost epiphanies). Their emissions caused the global oceans to acidify and temperatures on land to skyrocket by 10° to 30°C. Eons before cryptocurrencies or coal-fired power plants, and long before our ancient ancestors unlocked the secrets of fire, climate change nearly ended our story before it had even started. It's hard to conceptualize the scale of these losses and how profoundly they shaped the Earth, even on what feels like the cusp of our own apocalypse.

The past is not a perfect analog for the present, of course. The heat wave that triggered the Great Dying took around seven hundred thousand years to unfold, and ours has barely spanned the breadth of a human life (though we're fortunately not anywhere close in terms of the magnitude of warming of the end-Permian). If you're not an aficionado of deep time, it might as well have unfolded in a galaxy far, far away, this momentous turning point in the history of life buried in the layers of a past so distant we struggle to comprehend it. The story lies deep beneath our feet and far beyond the reaches of memory. But for those who can speak the language of fossils and sediments, the past is not just an object lesson; the Earth has left us a road map for how to survive the climate crisis.

For those of us born into individualistic cultures, the vastness of deep time can be just as terrifying as it is comforting; it provokes our deep-seated fear that we are insignificant and powerless, even as it assures us in our darkest moments that things will not always be as they are now. But just as an ocean is a multitude of drops, eternity is an amalgam of moments: the minutes, hours, and days in which we find ourselves bound together, and to the planet, with a charge to be good ancestors. In five billion years—nearly as long as the Earth has existed—our sun will explode, regardless of whether we are very good or very bad at tending to the planets it illuminates. This fact does not lessen our responsibility to fill that time with as much regard for life as we can, in the liminal space between Earth's creation and destruction. We have always known this, deep down; why else would we plant trees that we know will never shade us?

The great irony of the fossil record is that we wouldn't be here without extinction. Had the dinosaurs not died out, there would have been no age of mammals and no "us." This is a debt we cannot pay back, but we can pay it forward. My scientific training has prepared me for the climate crisis, but it's my humanity that compels me to do something about it. Earth's deep past is not a license to test the limits of ecological resilience, nor is it an assurance that, no matter what becomes of humanity, "the planet will be fine." The hard rock record does not promise that we, in all of our soft-bodied ephemeralness, could not possibly do as much damage as an asteroid: in the climate crisis, humans *are* the impact event, but we are also the small furry things emerging from the safety of our burrows in the aftermath and the ferns renewing the blasted landscape with greenery, creating something new out of the ashes of the old world. Unlike the dinosaurs, we have a choice: Will we be the asteroid or the fern?

What advice would a trilobite give us about the climate crisis? They thrived in Earth's oceans for two hundred and seventy million years, surviving two mass extinctions until they succumbed to the Great Dying. That's a long time to become wise. They were one of the most successful groups in evolutionary history, even though they left no direct descendants. Paleontologists attribute their success in large part to their exoskeletons: their shells were tough enough to survive the many dangers lurking at the bottom of the shallow Devonian seas. But we must be strong, not hard. To somehow protect the softness inside us without the false armor of nihilism or the nakedness of despair. Look around you: every living thing you see is a survivor. Not just of the climatic turmoil of the last ice age but of the countless cataclysms before it. How can we not be humbled by these lessons in resilience?

If not a trilobite, consider the muskox, the collective, matriarchal, stubborn, resilient beasts that they are. Across the tundra, their hooves drum across the bones of the mammoths they once grazed alongside. The pregnant females decide where and when the herd will roam every summer, adapting to the needs of the group and the capacity of the land. Muskox power through the long Arctic winters with a combination of resourcefulness and sheer will. When threatened, they form a circular phalanx around the vulnerable, their tough skulls and protective horns presenting a formidable defense. They are no strangers to loss: they have learned what it takes to survive.

What could we accomplish if we stood together and faced the danger? What seeds might we plant today that will one day take root above our bones? What if the future was better than the past? What if it was beautiful?

In Praise of
Indirect Consequences

Rebecca Solnit

I had a visit recently from Saket Soni of Resilience Force, and Marielena Hincapié, who had just stepped down as head of the National Immigration Law Center. We sat outside in my shared small backyard, squished between row houses and high fences, on a hot day when the shade of the fig tree was very welcome. A word about the fig tree: most years the tree doesn't get enough sun, and San Francisco just isn't hot enough for the figs to get really sweet; they're a bit watery and bland. But years after I took charge of watering and pruning it and got it to go from sapling to small but thriving tree, I realized that the fig tree was not for the three households in the building.

It was for the parents of my downstairs neighbor. They are working-class people who, in their youth, emigrated from a small town in Italy to Montreal, and they've pined for their Mediterranean climate ever since. The fact that their son has a fig tree brings them joy, and they ask about it on nearly every phone conversation, he tells me. The most valuable fruit this tree bears is intangible and consumed halfway across the continent. It's a fig tree bearing fully ripe and very sweet indirect consequences.

We sat under it, drinking homemade lemonade from a lemon tree I planted fifteen years ago, and I told them I'd heard that a lot of young people who organized to pass the Green New Deal felt defeated. Both my guests assured me that they thought the Sunrise Movement's impact was profound and in many ways a victory, even if not with any simple finish line. Resilience Force, if you don't know it, is about bringing together the undocumented laborers who clean up after climate disasters, to create solidarity and protect rights and recognize and cultivate their skills, and also to build recognition of their value in the communities they serve and build bridges between the residents and these workers. Like the Green New Deal, it recognizes the intersectionality of labor justice and climate justice.

Activism routinely consists of a movement, a manifesto, a group demanding something and not getting it, at least not at first. Too often, people seem to think that if there are not immediate and obvious consequences, there's failure. In reality, what happens in response is often more subtle, delayed, unpredictable, incremental, and indirect—and yet still valuable and significant, sometimes more so—than simple formulas and short timelines account for. Often those consequences continue to ripple outward and unfold for decades afterward. These impacts are not always positive—for example, the February 2022 Russian invasion of Ukraine led to food crises in other parts of the world. But sometimes they are, and they matter either way. As they do for the Green New Deal.

Later, I asked Saket to repeat what he'd said under the fig tree, and he emailed me, "The Green New Deal is the product of a prophetic imagination. It sees a world we all want to fight for and belong in. And like the best 'radical' demands, it's perfectly logical, even unsurprising, once it's out there. It's hard to imagine a time we didn't want it. Perhaps its most important contribution is the way it inspired us back into faith in large-scale, broad-shouldered collective action."

Maybe my favorite indirect consequences story in recent years is of how a station wagon full of twenty-somethings from New York City showed up at the Lakota people's Standing Rock in 2016. No one had heard of any of them but one of them was so galvanized by what happened there—

But before we get to that, let me note a few things about Standing Rock. The linear argument would be that it was a movement and an uprising to stop a pipeline, and the pipeline was not stopped. I believe that the permit could still be revoked or its illegality recognized some other way, and I remember when the doomsayers were telling us that we would never stop the Keystone XL Pipeline that would have shortened the route to bring dirty crude from Alberta to US refineries for sale overseas. In 2021, more than ten years on, the KXL campaign was finally victorious, but along the way, it built coalitions, educated people about the tar sands, and inspired people to campaign successfully against other pipelines. The process was no small part of the product.

It's unwise to rush to conclusions. In late 2021, Harvard and Princeton universities announced they would divest from fossil fuel. It took organizers ten years to make that happen. For more than nine years, you could have looked at the campaigns as not successful, even though they were part of a global movement that got trillions of dollars out of fossil-fuel investments and raised ethical questions for all individual and institutional investors to consider. Sometimes change happens slowly. Often, when it happens suddenly or appears to do so, it's because the consequences are suddenly arriving for longtime change or organizing that previously seemed inconsequential, or the big change is the visible public effect of not-so-visible work. And it does sometimes come suddenly, as earthquakes and watersheds, ruptures from how things were before.

Standing Rock itself—so much happened there in 2016: the broadest gathering of Indigenous nations maybe in the continent's history; the first convening of all seven Lakota nations since the nineteenth century; hundreds of veterans getting down on their knees to apologize for the US army's genocide; alliances and friendships built. You can't measure what these things did, but they mattered.

A Standing Rock organizer I worked with on pandemic medical relief told me that the gathering and resistance gave young people there a sense of hope and agency and of their own value that they did not have before, and a Utah climate organizer told me the same thing was more broadly true of the young Native activists he was meeting. A lot of non-native people got an education in Native American history and how it intersects with environmental and climate issues.

So, about that station wagon—one of the people in it was a Bronx-based bartender of no particular fame at the time, but that description might already clue you into the fact that, moved and inspired by Standing Rock, she'd run for Congress in 2018—"I first started considering running for Congress, actually, at Standing Rock in North Dakota." she said. "It was really from that crucible of activism where I saw people putting their lives on the line . . . for people they've never met and never known. When I saw that, I knew that I had to do something more."

And that brings up the achievement of the Justice Democrats who supported the successful campaign that made her, two years after that journey to South Dakota, the youngest woman ever to serve in Congress. Like the confluences of streams that make a mighty river, there are confluences of organizing and ideas and commitments. It pained me that so many people regarded Bernie Sanders's 2016 campaign as a bitter defeat rather than a resounding victory in proving there was a constituency and momentum for progressive ideas that could be built upon in the years to come. The Justice Democrats came out of that campaign, and that congressional victory is a credit to their vision and their understanding of the machinery of change. I know Sunrise worked with them, and that the election to Congress of Rashida Tlaib, Ilhan Omar, and Deb Haaland (now, as secretary of the Interior, our first Native American cabinet secretary) is also connected to their good work.

As for the station-wagon traveler: of course, you know I'm talking about Alexandria Ocasio-Cortez, who would become a hugely visible and audible figure and a significant voice for progressive issues. She sponsored the Green New Deal in the House of Representatives in February 2019, little more than a year after taking her oath of office. Notice the confluence of Standing Rock, the Justice Democrats, Sunrise, and this young woman's life. She said, "I was really wallowing in despair for a while: What do I do? Is this my life? Just showing up, working, knowing that things are so difficult, then going home and doing it again. And I think what was profoundly liberating was engaging in my first action—when I went to Standing Rock, in the Dakotas, to fight against a fracking pipeline. It seemed impossible at the time. It was just normal people, showing up, just standing on the land to prevent this pipeline from going through. And it made me feel extremely powerful, even though we had nothing, materially—just the act of standing up to some of the most powerful corporations in the world. From there I learned that hope is not something that you have. Hope is something that you create, with your actions. Hope is something you have to manifest into the world, and once one person has hope, it can be contagious. Other people start acting in a way that has more hope."

She went from looking for hope to making it, through her work on many issues and her brilliant leadership on key issues for the country and the world, including climate. The Green New Deal, longtime activists agree, changed the whole conversation about what is possible and what we want. It broke down the false binary of jobs versus the environment, which had bedeviled the environmental movement for decades. It scaled the ambition way up on what we can and must do, and how all the pieces can and must fit together. It was a jobs-and-infrastructure program, and a farm program, and a justice program. (It was a proposal mostly about carrots, not sticks; it didn't take on the fossil-fuel industry directly, as many climate groups and campaigns have.)

It provided a template and momentum for regional and international climate plans and proposals, including Ithaca's citywide Green New Deal, and both the Global Green New Deal launched, in 2019, at an international mayors' conference in Copenhagen and the European Green New Deal. The GND was a template for Biden's climate platform and then for Build Back Better, which in its boldest and most visionary parts *was* the Green New Deal.

Another activist success in changing the conversation, by imperceptible increments, is that nearly every Democrat in the House and Senate was willing to support legislation—in the form of Build Back Better—that only a few years ago had been considered alarmingly radical. And then Build Back Better got beaten down by one vote—from the coal-baron senator Joe Manchin—but came back as the Inflation Reduction Act, also addressing health care and taxes and dedicating $369 billion in spending on climate-related matters. Again, what was best in it had its roots in the Green New Deal. It was not a perfect or comprehensive victory, but it was a victory (and what was worst in it was fought, with some success). What was in it for the climate would have seemed impossibly ambitious not long before.

Daniel Hunter, an organizer with 350.org and the Sunrise Movement, wrote, after the Inflation Reduction Act passed, "Movements don't ever win. At least, not in the sense that most of us typically think about." He means that the ultimate victory doesn't come to pass. You become an activist because of an ideal, and that ideal, if it is going to be realized at all, will usually be realized only imperfectly and partially. So, "there's no moment when we see our values are suddenly acted upon by all." Nevertheless, he argues, we need to celebrate. I've often noticed that people fear that claiming victory means believing the work is finished, but Hunter argues that, instead, it helps us keep going. Claim victory, he argues, "because naming and reviewing the movement's progress helps build momentum for the next win and the win after that. People want to join positive teams that burst with love and energy. And this is where celebrating is so crucial."

Something I once learned about fame might be true, in another way, of ideas. A moderately famous artist is recognizable and visible, but a truly impactful artist is no longer *what* you see but *how* you see. They have become invisible because they are behind your eyes rather than in front of them. They are how you see everything else, not just their own productions. The Green New Deal is not held up as much now as a good idea we should pay attention to, because it has just become how we think about renewables, jobs, climate, and justice.

The climate crisis is an emergency. We need some very direct action and some very big change, and we need it fast. If you look at all the small pieces—this wind farm launched, this pipeline defeated, these trillions divested, this public engaged, this election won, these measures passed—it adds up to a lot. The struggle is long, and boundaries are important. Still, a huge shift from where we were a decade ago, in terms of broadly shared understandings and agendas, the improvement and implementation of renewable energy and electrification, the growth of the movement not just in scale but in sophistication and intersectionality.

What we aspire to seems hard to reach, but the improbable, for better or worse, is not the impossible. An essay on activist leadership taught me, "Moses Maimonides, the Jewish scholar of the twelfth century, argued that hope is belief in the 'plausibility of the possible' as opposed to the 'necessity of the probable.' While it is always 'probable' that Goliath will win, it is also true that sometimes David wins, a sense of the 'possible' that we experience in our own lives as well. Hope emerges from this sense of possibility, freeing us from the shackles of probability."

The improbable is our job. We have to imagine a world in which the Green New Deal is just the way things are—or in which it has been superseded by greener deals. I'm hopeful. And I'm grateful for what the young people of the Sunrise Movement have set in motion. This story is not finished, and we do not know how it ends. But we can help decide that. Doing the work matters. Knowing how and why it matters, why it's worth it, means including its indirect consequences.

From the Hunger Strike with Love

Nikayla Jefferson

I often wish I could forget my twenty-fifth birthday. The memories of that day are some of the most unforgettable yet of this life. These memories are still fresh, and at the time of this writing, I'm still unraveling their intense emotions. These words that follow are the best I can offer about how it felt to organize the Hunger Strike for Climate Justice in the fall of 2021.

They starved. I emailed details to the press. I posted photos on Instagram. I edited videos for our Twitter. Their bodies deteriorated into a state too weak to walk to their little red chairs outside the tall black gates of the White House, so we each took turns wheeling them instead. They starved. Their bodies were too tired to talk to visitors or journalists, so we unfolded cots and laid sleeping bags away from the sun. They starved, and I organized their starvation. My job for this action was to support them through their choice, tell the story to the world, and make their sacrifice matter in ne-gotiations over Build Back Better. They starved, and I cried at the microphone for my opening remarks.

Paul, Ema, Julie, Abby, and Kidus starved for twelve days because the *New York Times* broke the news that the climate policies necessary to avert planetary disaster were erased from President Biden's Build Back Better agenda. For the past few years, we all worked as organizers in the Sunrise Movement, alongside the wider climate movement, to win the public and political support to make this moment with Build Back Better possible. We all believed our best chance against Earth's worst-case scenario was a federally funded, nationally coordinated, decade-long mobilization against the climate crisis—and this was the critical moment to jump-start our best chance. The year 2020 marked the end of the hottest decade, elected the party of climate action into total government power, and saw the majority of Democrats in Congress now ready to pass the Biden administration's Green New Deal–lite. We all believed this moment must be the beginning of the United States' long-overdue climate action.

Kidus and I hatched a plan, and we called a few movement comrades to ask if they, too, felt called at this moment. I borrowed the title of a *Time* magazine article from the previous year about the imperative of the moment our political stars align for our Sunrise-wide Zoom announcement of the hunger strike: "The Last Best Chance for Humanity."

The urgency felt unbearable, our belief felt nonnegotiable. Because without Build Back Better, Earth's worst-case scenario felt imminent. The truth of the moment rang out from the bottom of my heart and swelled my despair to an all-time high. I could not unsee the vision of this version of the universe: the pained cries of hundreds of millions of people suffering in famine and drought, swaths of Earth no longer fit to nurture life, and the life that remains struggles on through the fires and floods. I could not unsee a vision of this version of California: it burns all four seasons, faucets run dry, and my family is forced to migrate because the ocean swallows our home. The version of the universe in which Earth warms past its liveable limit, because Joe Manchin refused to pass Build Back Better, felt in that moment unsurvivable. Even if not because of the climate, I felt would die in this version of the universe because my heart would be so stricken with grief. The moment really did feel like "The Last Best Chance for Humanity."

I was not alone in my feelings. We announced on Monday, and on Wednesday morning I tweeted out a photo of Paul, Ema, Julie, Abby, and Kidus in their little red chairs outside the White House gates, holding their "I'm on Hunger Strike for Climate Justice" cardboard signs. It was October 20. Day one.

October 24. Day four. Kidus could not stand up. His hands were numb. He was hot all over. Roberto, another member of the support team, carried him into the back of my rental SUV. I drove fast to the hospital. Kidus was clear with me before he started the hunger strike: "Record everything." I did my job. Slumped over, slurring in my backseat. Eyes rolled back and unresponsive in the emergency waiting room. Sunrise shirt pushed up to give the nurse a bare chest for an EKG. In a private room while we waited for the attending doctor, his heart beeped on in the quiet. I wondered in this brief moment of stillness, *Will it really come to this boy dying for the Democrats to just do their fucking job?* I cried in the quiet because the answer felt like, *It just might be so.*

He woke up with a lazy smile. "Did you get good pictures?" *Of course.*

The nurses came back in, asked him why he had not eaten in four days and refused their hospital food. "Have you heard of Build Back Better?" he asked them. They hadn't.

Kidus was born in Ethiopia. He lives now in Texas with his family. He wonders, worries just like I do, about how the climate crisis might come for his people and his homes in the future, how it comes for them now. He despairs about it just as I do, and it is the reason he sleeps in this hospital bed in total starvation. The *why* of his choice to hunger strike is the same *why* of my choice to organize it and the same *why* of this moment I sit in this room beside him: because we love.

Kidus returned the next morning to the gates outside the White House.

October 25. Day five. I drove fast and far through the midnight and out of the DC limits. I played *Tiny Dancer* at full volume, windows down, flying on an empty highway because I wanted to slip away into a version of the universe more like 1976, a time I imagined I would feel young and carefree on this midnight drive for a spontaneous reunion with a lover. A triple-scoop Baskin-Robbins waffle cone dripped on my thighs. I drove fast and far because I was twenty-five on the twenty-fifth, and by the rules of "golden birthday," this day is meant to be the best one of my life. For just a few early morning hours, I wanted to slip away into a place not here and a time not now for the chance to just pretend.

The hunger strike ended on day twelve, a week after my birthday. The health of the hunger strikers was in serious decline, and the support team and their families were in a proportionate level of acute distress. Joe Manchin was now even more unlikely to move from his "no" position, and we doubted our power to shift his vote at this moment. The hunger strikers truly gave "The Last Best Chance for Humanity" the whole of their minds, bodies, and spirits. We received a stern warning from our medical team that any more could result in permanent disability, and unexpected complications could be their death.

The hunger strike ended for the public in one of those wild and improbable twists: the hunger strikers and a hundred friends blocking Joe Manchin in a parking garage, a video of him behind the wheel of his Maserati, revving his engine through the crowd. "Maserati Manchin" trended #1 on Twitter that day, and the video went viral. It was not federal climate policy, but it felt we were gifted a small consolation prize.

In the waning days of the hunger strike, my mental health spiraled. After we packed up the little red chairs on the last day outside the White House gates, I wanted nothing more than my parents. I returned home to San Diego for solitude beside the ocean, and there I contemplated the *why* of it all. I felt proud it happened, that it mattered, and that I was part of it. I felt grief that it happened, that we had to do something so dramatic and sacrificial to matter, and that parts of me were traumatized in the process. I resented the *why* of it all: love.

The truth of my climate despair is that it is a tender ache around the space of my heart, but I feel this great pain only because I feel greater love. The despair is a consequence of the many wells of my heart filled with love for family and friends, for my California and the land, for all the life, human and otherwise, that colors Earth with beauty and gives me enough meaning on darkest of days to endure on. The despair is the heartache of loss, the fear of more loss, and the empathy for all the other life living this moment of loss alongside me. I organized the hunger strike for Climate Justice because this despair does not close off my heart into apathy or strike it into paralysis with grief—it is the fire that burns my heart into action. Love and its impulse led me to spend my twenty-fifth birthday organizing a hunger strike. It was not how I imagined my golden birthday, looking from my younger years into the future, but I do not wish it happened any other way. I believe it happened just as it was meant. The choice led me here to this essay to you.

Full Narratives
of Love and Hope

Interview with Fenton Lutunatabua and Joseph Zane Sikulu,
Pacific Climate Warriors

The Pacific Climate Warriors are a youth-led grassroots network working with communities to fight climate change in the Pacific Islands. The Pacific Climate Warriors work with organizers across seventeen Pacific Island nations and diaspora communities in Australia, New Zealand, and the United States.

Thelma: My first question is about the maxim of the Pacific Climate Warriors: *We are not drowning. We are fighting.* Can you tell me where it came from and how it's guided your work?

Fenton: In 2013, about fifty Pacific islanders gathered at the Global Power Shift, which was a massive gathering of about five hundred young climate activists from around the world. And this configuration of Pacific Islanders had never met together collectively before. We had heard of each other, working in different climate spaces and social justice issue spaces. When we gathered and started talking story, we realized that so many of the ways in which participants at other global gatherings thought about Pacific Islanders was in a very condescending way, very like "poor brown people."

And then we realized that was largely informed by this narrative, of Pacific peoples and folks in the Global South, that we were mere victims of climate change and needed to be saved. We were recipients of a white savior narrative that really painted us as folks who really couldn't tap into our own agency or, you know, into our innovativeness. So, as that configuration of people, we knew there needs to be a more nuanced story that is told about our people, and we really need to claim ourselves as the heroes in these stories. We started thinking about what that looked like. There are so many beautiful things about warriorship and stewardship from the islands that we want to uplift. And so, we're just like, let's just keep this simple and go with Pacific Climate Warriors. *We are not drowning. We are fighting.* It was a simple sentence that became an anti-narrative to the dominant story told about a people. All of these stories were being told about us without us, and it really was a moment of claiming back our full identities and full narratives.

Thelma: Joe, how has it guided the work in the years since then?

Joseph: I think it's grown up to become a real rallying cry. Whenever anybody thinks of us, those are the words that ring, and they've rung out in so many places, and it's been incredible to see it just not just chanted at rallies but to hear our leaders take on that mantra, to see it flowing out in communications all over the world. But also understanding that there's been some real intentional strategic thinking around how we use that messaging.

It's become such an empowering cry, especially for young people in the diaspora who have grown away from home. They can find a space and meaning behind words where they can actually practice their culture in its entirety and share understanding of that to bring about change.

Thelma: What do you think is the Pacific Climate Warriors' approach to staying mentally, spiritually, emotionally strong while tackling the climate crisis? Especially with so many of the harsh realities in the Pacific.

Joseph: Well, I think that one is quite simple in that our work is grounded in love and care, and we take the time out to ensure that we invest in each other, in our well-being, in our health, in our mental health, in our faith. That we are grounded in our love for each other and our relationships, and that we're always holding each other.

What I love about the Pacific Climate Warriors is that we spend so much time building a container around our work in the spaces that we're in, in the relationships that we hold. Everybody knows where they're at and that they're going to be held throughout. It doesn't matter what kind of adversity we face, whether it's the climate crisis or in our personal issues, or whether we're carrying so many different things in our daily lives—you walk into a space and you know you're always going to be held in love. Love is, I think, one of the most underrated organizing tools, because people will just take it for granted, but it's what brings everybody back into this. One of the cool things of being a Pacific person is that everything always starts and ends from a place of love.

Fenton: Yeah, I think love, hope, community. Everything is so relational, right? This intentional investment and commitment to relationality and having connection at the core. I say that in not a fluffy way, but in the world that we're building, in the world that we're creating and the things that we're handing off to the next generation—if love, hope, and community aren't at the core of it, we're not doing this right. So much of capitalism and colonization seeks to destroy all these things in the name of profit. I think in the building out of this new world, we have to, now more than ever, fight for the things that we love, the lands that we love, the place that we love. You know—the people and the islands that we call home, and I think a lot of that means being in the right relationship with both our communities as well as the natural world.

Thelma: Something I've noticed about the Pacific Climate Warriors spaces is how joyous they are! For you, what role does creativity, joy, culture, fashion play in your work?

Fenton: Yeah, I think so much of it is about meeting people where they are and saying, come as your full self. Within the Pacific Climate Warriors, we have people that are designers, that are stylists, accountants, lawyers, artists, poets. And what we do is truly try and make it as accessible as possible for people, and just make sure when they do show up, they can bring all of their gifts, and identify ways in which they can use those gifts to serve this group of people, this moment, this time.

So much of the way we've been intentional about storytelling through images and films has been about, *How do we tell our own stories that are not rooted in whiteness?* This is our standard of what we believe beautiful is, this is our standard of what we believe warriorship is. It is so completely disconnected from the white climate movement. It truly is ours.

Joseph: We have worked really hard to build our brand of activism and try to shift the way the world looks at what an environmental and climate activist looks like. We love our island prints, we love our seis, we love our flowers. Those are also just a representation of who we are, but activists, especially Pacific activists, can look like anything. There's no one set way in which you're supposed to look. It's really just about embracing everything that you are, because that's what we're fighting for. I think our brand of activism has also been what has drawn so many people into it, because we just stand fully in our identity and culture. And there's nothing wrong with climate activists looking really great! We look amazing, and I think that the world we want to build is a beautiful world, and one way to show that is by projecting everything that we're trying to fight for.

Fenton: I also think so much of the ways that we tended to our bodies was looked down on and frowned on. Growing out my Afro is like a form of resistance for me, because I grew up my entire life saying I needed to have a buzz cut, because that is the disciplined way to look. So much of the ways in which old people and our ancestors moved through this world, all of that was seen as not good enough. So, it's that intentional pushback. When we talk about fashion in the Pacific, it's not a bougie thing. Many of our designers and our stylists operate with nothing. One of the things Joe and I talk about all the time is creativity and scarcity having this beautiful relationship. Many of the designers' prints tell stories. It comes from stories that they grew up with from their grandmothers, from the grandparents. It tells stories about the natural world, the ocean, and to be able to carry that as like armor into a UNFCCC [United Nations Framework Convention on Climate Change] space, you know we're representing. So many times, our voices have been silenced. This is something that we still battle with. So much imposter identity, because we've always been told we don't belong in that space, and so if we are in this space where we can't use our voice for whatever reason, we'll use our bodies to tell that story.

Thelma: What would you say to those who are feeling overwhelmed by the state of the world and uncertain about how to get going on their climate journey?

Joseph: Drink some kava. (*Laughs.*) Yeah, I think, know that you're not alone—we all feel overwhelmed sometimes. Our way through that is by finding connection. We don't have to go through it alone. Find connection and find love, and sometimes it takes a bit of work to find that community that you're looking for. Reach out, just take a chance and step forward. We need to push ourselves out of our comfort zone sometimes to do that, because you never know what we're going to be on the other side.

Fenton: You do not have to do this on your own. So much of this story about individualism needs to be left behind. The future needs to be one that's collective and communal.

THE FUTURE
WE WANT

This is not the way things have to be.

—**Angela Davis,**
quoting her mother's words to her
when she was young

There's a story that begins here, or maybe it ends.
It depends on us.

—**Robin Wall Kimmerer**

We live in capitalism, its power seems inescapable—but then, so did the divine right of kings. Any human power can be resisted and changed by human beings.

—**Ursula K. Le Guin**

As this century draws to a close, a century packed with history, what leaps out from that history is its utter unpredictability.

—**Howard Zinn (1988)**

Imagination Is a Muscle

A Conversation with adrienne maree brown

Thelma: I love your collection of visionary futurism, *Octavia's Brood*. The introduction, written with your coauthor, Walidah Imarisha, says, "Once the imagination is unshackled, liberation is limitless." If we, as a society, were able to unshackle our imaginations, what could be possible?

adrienne: If we were able to unshackle our imaginations in this moment, I think our compatibility with the Earth would become possible. I believe that humans have a concept of ourselves as taking from the Earth and not necessarily giving anything back to it, using the Earth as kind of a temporary machine from which we're going to launch some other situation or find some other planet. And it's a very limited worldview. I think if we were to unshackle our imaginations, we would be able to see this is an abundant place that has everything we need. It could actually be very satisfying for us. Maybe we could satisfy it, too. When I imagine the world in a right relationship, it's a love story between our species and this Earth and amongst us.

Thelma: What are your suggestions for people who want to kick-start their imaginative power?

adrienne: Imagination is a muscle that, for many of us, will atrophy if we don't use it, especially under the pressure of constant fear. Fear and imagination often can't be in the same room. So, one of the things I think to strengthen that muscle is, first of all, reading. Reading visionary fiction, reading visionary texts. I often recommend Octavia Butler to people because it's reading things that are hard. Walidah Imarisha always says, "It's realistic and hard, but it's hopeful that change is possible." How do we be with what is and keep our eyes up? And then I love collaborative science fiction writing and collaborative visionary fiction writing. Sitting down in a circle, identifying what is something in our community that needs the medicine of imagination brought to it, and generating ideas together.

Rebecca: I think of the moment we're living in now as a kind of wild science fiction unimaginable years ago. What are some of the things you've already seen to come to pass? The kind of big changes produced by the kind of social murmurations and emergent communities.

adrienne: Some of the things that have most astounded me that have come to pass in our recent history have been the ways we have galvanized and changed the culture and conversation around sexual harassment, harm, assault, and violence through the #MeToo movement—and understanding that that change was made possible because of a lot of people at a relatively small scale being willing to tell their stories, tell their truths, and begin to make interventions for themselves by stepping out into the light. Concurrently, we've seen a major shift in the culture and the conversation around abolition and prison systems and the preciousness of Black life through the work of Black Lives Matter and the Movement for Black Lives. And with both movements there have been massive learning edges.

Yet, if I think about ten years ago, that didn't feel possible, it didn't feel like that could happen that quickly. And that gives me the most hope, because I really do believe there's some almost intangible impulse that moves in us. There's something that will click after, sometimes, generations. I'm the descendant of people who were enslaved, so I know that there were generations where people were enslaved, and it was just like, "This will never change." At the same time, there were those who were pushing and pushing for it to change all along. So then, this impulse happens. It's like, "Okay, now is the time." Which we see in nature often. It's like, "Oh, the birds are all kicking it, and now they've all taken off." What happened, right? We're just like that. We're available when the pressures are aligned and there's an opening—we will fly through it. I think the same is about to happen for us with climate. A component that's necessary is there has to be something that people can do at an individual level that feels meaningful, and that combines with having a large-scale analysis of what's going on that is collective.

Thelma: I'm so glad you brought up nature as well, and you speak to nature as a teacher in so much of your work. What are some solutions from nature that you think we should bring forward into our climate work?

adrienne: Nature has been one of the most important teachers of my life, and I'm just at the place where I'm started to really understand that humans are fully a part of that. For the longest time it was like—there's nature over yonder, and there's humans over here. But now, I realize the reason nature can teach us is it's us. We are that, and that is us. You know, we are the deltas. We are the river or the mountains. Lately, I've been thinking of humans as just another form of water. There's rivers, there's raindrops, there's humans. We're just a different structure for water, and one of the first lessons is water is a mutable substance. Water is always changing from one form to another, given whatever task is needed. Right now, we have to be willing to become kind of a vapor. We have to be able to be really light and be able to rise up much higher than we imagine even being possible. That vapor rises and it gathers, and it becomes rain, and when rain comes back down sometimes that's the only thing that can take out a wildfire. I keep thinking the move we need to learn is, how do we be light enough to gather ourselves up into something massive, and then heavy enough to really fall and be a monsoon against these wildfires that are happening amongst our chance at life.

Rebecca: I love that you're finding these wonderful survival metaphors in the natural world. And I wonder how you can speak to the climate crisis as a storytelling crisis.

adrienne: Probably the reason we're in the climate crisis is because we're living inside a story of accumulation. Somewhere along the line there's a story of scarcity, as if this planet is a scarce planet, and we must compete to accumulate as much as we can. And that story has such a strong grip on us—that generations will never have enough but constantly must be striving for enough in ways that support the very top-tier percentage. The story—it's a glitch in this system. As a result, I am constantly astounded by how people are organized to move against their own interests by that story.

When I think about what we need to do related to climate catastrophe and climate change, the stories I think we need to tell are ones of what it is like to be in a relationship with the Earth. What is it like to live on an abundant Earth? What are communities that are thriving by being in relationship to the Earth? What does enough feel like?

We have this idea that there's scarcity and shortage. But if you've ever put a seed into the ground, you see what a fecund world we live in. It's over-providing. Every time I've tried any gardening experiment, I'm like, "Okay, now there's too much of everything," and we must jar it and try to distribute it among everyone. That's the nature of the Earth that we live on. That's the story that we want to tell. And if there's places that are not experiencing that abundance, that is because of human impact, not because of the nature of the planet itself.

Thelma: Like what you said, so much of the narrative is around scarcity and limitations and denial. And that extends to activism, too—to be a good climate activist, you have to be sad and angry. You are just such a master when talking about pleasure. For you, what is the connection between pleasure and hope?

adrienne: I think you have to be dosing yourself with pleasure to maintain hope. All along your life you must find the small pleasures that keep you in touch with what life actually feels like, and help you slip under the lie that we're in a total crisis and there's nothing good happening.

One of the stories I'll say is there's a community here in Durham, and I just got to go visit them. It's called Earthseed—it's named after this community in Octavia Butler's *Parable* books. I pulled up for the first time to this space, and it was a delight. It's a garden in the middle of the summer in North Carolina, out in the sun. We're sitting there, they've got fans blowing on us, while these gorgeous queer Black and brown teachers taught us about herbs that we could use to heal ourselves to make tea, and how available they were. There was so much wisdom. And then these children came running over with frozen bags of fresh strawberries that they had picked from that garden. They then made these strawberry smoothies, and you could taste the sun in your mouth. It was the most delicious, sensual, delightful gift, and they were so proud. The kids were so proud of themselves to be able to offer up this strawberry gift to us. The seven-year-old knows the magic of strawberries. We're going to be fine.

Even as there are people training their children to operate against humanity, or to operate against certain parts of humanity, there are these other miraculous babies who are being raised to love each other and love the Earth, and that is delightful. Kids are the experts of pleasure.

As adults I think it's also important, especially on the hard days, it feels even more important to go outside and put your feet down in the Earth and ask her to hold you right and ask her for wisdom and guidance. I watch all the sunsets, and I make sure that the generation coming after me knows how important sunsets are, and corn and bananas and coconut water and the kale out of the garden.

Rebecca: One thing I found helpful and profound in your thinking is your response to negative emotions, to fear, to grief, and so on. A lot of times, people think these are poisons. You see them as something that can connect us and deepen our relationships and awareness. Do you want to talk about that?

adrienne: Well, I think a lot of us are raised to be oriented toward the joyful emotions, the happy, uplifting, useful, and even the numb emotions. We're not given coherence with the hard emotions of fear and disgust and anger. So, when they come, we don't know what to do with them. We might try to repress them, or they might explode out of us and/or overwhelm us. I think a lot of that leads to isolation, because we either act out in a way that people isolate us, or we pull away to feel our hard feelings.

When I'm grieving. I know that I'm not grieving because death is unnatural, I'm grieving because love has overfilled my banks, and I can no longer just pour it into this other person. When I'm angry, it's because I love something, and I want to defend it. When I am feeling fearful is because I love something.

Particularly around the climate right now, if you don't let the despair flow through, and let it in and grieve about what we have lost—if we don't let that come, it's going to find other ways to surface itself. I keep telling people, sometimes I'm crying in the shower because I love the Earth so much, and I want us all to put it first. And then I let that despair focus my writing and focus my offer in the world. I want my lineage to be a lineage of someone who fought hard for human existence on this planet. My despair and grief and fear and anger informed that, and they're powerful parts of that fight, just as much as my love.

Looking Forward from the Past

2023 FROM 1973

Rebecca Solnit

In 1973, the present stretched, as it usually does, to the far corners of awareness and was easy to mistake for forever. What was ugly, painful, unjust seemed as though it would last forever or for the indefinite amount of time that is easy to mistake for forever. The same was true of what seemed safe, stable, and good. Which is to say that 2023 was, if not unimaginable in 1973, at least unimagined. Nineteen seventy-three was a turbulent year, one in which many found it easy to be cynical or despondent.

Still, people believed they could see what was crumbling, fading, or imploding. In the US, the civil rights movement was widely regarded having run aground in 1973. But, for Native Americans, queer people, people with disabilities, and women, new rights movements were gathering momentum, and their biggest victories would happen in the 1990s and in the new century. "Once in a lifetime," wrote the poet Seamus Heaney, "justice can rise up / And hope and history rhyme."

On October 14 of that year, a Black baby was born in Fayette-ville, North Carolina, whose murder by police on May 25, 2020, would set off the biggest protests in US history, building on a new wave of civil rights movements around race, policing, prison, and economic justice, drawing in first-time participants, changing the national conversation. His name was George Floyd, but no one knew he had a place in history until he was murdered in public by a police officer. Every era has its people who are invisible, unrecog-nized, ordinary until they are not; we can be confident that they are among us now and pray that their impact comes through their lives and not their deaths.

In 1973, as in almost any time, few could see what was begin-ning. The way tiny seeds germinate underground, that ideas travel from margin to center, from outraged reaction to widespread accep-tance, from rights campaign to established law, the way germinating seeds would crack the surfaces that seemed solid and stable. The way that things as fundamental as food, nature, race, gender, spirituality, justice would be profoundly rethought, that equality and kindness would become more compelling values in more arenas. Inequalities hardly perceived then would become delegitimized (which invites the question of what we, in 2023, don't see that future eras will).

Politicians and a lot of ordinary people assumed the future would be some kind of increase, improvement, or decline of what was clearly visible in the present. But the invisible and overlooked always matter. Take population. In 1968, the Sierra Club had pub-lished an ugly, alarmist book called *The Population Bomb* that as-sumed the current birth rate would continue indefinitely, producing a huge increase in the number of human beings on Earth, predict-ing crises, including huge famines, in the 1970s and 1980s. The book, which described the people of the Global South as repellent, overabundant, and the source of the problem, was a success whose racist frameworks, though since repudiated by the Sierra Club, are still too widely believed today.

The Population Bomb stands now as a monument to failure of the imagination. It failed to comprehend its own biased assumption that the problem is those poor people over there rather than us rich people here, framing the problem as one of population rather than consumption. It failed to recognize that the fundamentals themselves—birth rates and the status of women—would change dramatically in coming decades. Global population growth peaked in the early 1960s and has halved since then; it was already declining when the book was published. Too, as Oxfam reported in 2020, "The richest one percent of the world's population are responsible for more than twice as much carbon pollution as the 3.1 billion people who made up the poorest half of humanity during a critical twenty-five-year period of unprecedented emissions growth." Consumption, not population, drives climate chaos and other environmental problems.

To a significant extent, the dramatic drop in birth rates was due to feminism—the ability of women to access reproductive education, care, and rights, and expand rights to social, political, and economic self-determination. The ability to control fertility—through access to reproductive education and health care, to choose when and whether to marry and procreate—is crucial to women's ability to function as free and equal participants in public and professional life. Over the next fifty years, women in many parts of the world would gain reproductive rights, including, just a few years before 2023, the right to abortion in Argentina, Ireland, Mexico, and Colombia.

In 1973, the Soviet Union was more than half a century old and seemed as though it would persist for decades or centuries more, but its collapse was eighteen years away. The Berlin Wall was twelve years old, and it would fall sixteen years later, in 1989, when the

Eastern European countries under the USSR's thrall liberated themselves in one of the greatest nonviolent civil resistance upheavals in human history. The cold war, the standoff between the nuclear powers of the USSR and what was then called the West, was also seen as a stalemate that would continue indefinitely. Then it all collapsed. No one foresaw how suddenly and completely it would all change. The world of 1991, let alone 2023, was unimaginable in 1973.

But the world was at least as turbulent as it always was. The 1973 oil embargo by Arab members of OPEC, the Organization of the Petroleum Exporting Countries, created an energy crisis for the United States and other countries dependent on Middle East petroleum, and it would drive the passing of legislation on energy alternatives, including research and development of solar, in the years to come. (The 1973 oil crisis in the US has its echoes in the 2022 gas crisis in Europe thanks to Russia's war games.) The famous 1972 *Limits to Growth* report from scientists at the Massachusetts Institute of Technology recognized the possibilities of many scenarios for population increase and was prescient about the threat to the climate of carbon dioxide emissions, but it could not imagine any alternative but nuclear power to the burning of fossil fuel.

Until the end of the century, renewables seemed like a minor part of the energy landscape, and the engineering and economic breakthroughs that would make a global transition away from the age of fossil fuel had yet to happen, and few imagined they would. Nearly every prediction about the cost of solar fell far short of its precipitous price drops; nearly every assumption about the limits of renewables was overcome by those invisible heroes of the climate movement, the engineers and inventors whose work continues to transform what is possible.

On September 11, 1973, a US-backed military coup overthrew Chile's three-year-old left-wing Allende regime and instituted fifteen years of authoritarian rule, with torture, disappearances, and murder of dissidents and the outspoken. Alberto Bachelet, a brigadier general who resisted the coup, was arrested that day and tortured until he died in 1974, and his wife and daughter were also rounded up and tortured. From the perspective of 1973, it would have been inconceivable that the daughter, Michelle Bachelet, would become the left-wing president of a democratic Chile in 2006 and again in 2014.

The neoliberal policies and economic inequality established by the Pinochet regime in Chile prompted huge student demonstrations in 2011 and 2019. A young Patagonian, Gabriel Boric, not yet born when Pinochet's coup took place, rose to prominence in the student movement. In 2022, he became Chile's president at thirty-five and appointed the first female-majority cabinet in the Americas.

President Allende is said to have committed suicide, though some believe he was murdered. His daughter Beatriz Allende was with him in the presidential palace that day. She and her young daughter, not yet two years old at the time of the coup, fled to Cuba, where she would die by suicide four years later. That daughter returned to Chile as a young woman; in 2022, Boric appointed Maya Alejandra Fernandez Allende head of the military that deposed her grandfather and drove her family into exile. It is one of history's most perfect about-faces.

But to tell it this way is to tell it as the story of high-profile individuals, whereas it should be told as the story of movements, of public participation, of grassroots organizing, of people showing up, often in the face of threats of bodily harm or even death, to stand on principle. They stood up as did las Madres de Plaza de Mayo in Argentina when it seemed impossible to topple the regime; they stood up when the risks were huge; they stood up to sing the *Ode to Joy* where political prisoners could hear their voices; they stood up to orchestrate voting in the 1988 plebescite that established an open election in 1990; they stood up to get out the vote that year; they kept standing up, showing up, speaking up. These countless, anonymous heroes were the midwives who birthed a new Chile.

As I write, the ideas and possibilities are opening up further. Though Chileans did not, in September 2022, vote to approve the new constitution, the fact that it enshrined the rights of nature and of Indigenous peoples signified a profound transformation, if one still in its early stages. These ideas were not on the table twenty or fifty years before. The present is always a place that was unimaginable at some point in the past. If 2023 was unimaginable in 1973, so was 1973 in 1923: this can equip us to recognize that the unimaginability of 2073 does not prove it is impossible. From that perspective, we are always living in a wild future, a future sometimes inconceivably better, sometimes unimaginably worse, usually both, from the vantage point of even the recent past.

And it's a reminder that these futures that become the present were made by our actions and inactions. At best by notable individuals, but also by communities, movements, newly framed constituencies, by public engagement, by people who worked toward what seemed at first impossible, then unlikely, and then one day became the order of things. By ideas that appeared small, on the margins, like mammals in the age of dinosaurs, like seedlings on the forest floor, like misfit children, and grew to become mighty themselves.

Looking Back from the Future

2023 FROM 2073

Denali Sai Nalamalapu

It's 97 degrees today. Nana and I are celebrating the break in extreme heat with a pitcher of sweet tea on her porch. Ma dropped me off on her way to work at the solar farm. My school flooded again early this week because of the heavy rains we've been getting. They'd barely finished repairing the art and design room and the cafeteria kitchen from the last flood, so they closed temporarily to get all the work done. That means I'm spending the day with Nana, getting a different kind of education, as Ma likes to say.

Nana and I sit quietly together, her rocking rhythmically in her chair, quietly humming to herself, me perched on her fading white front steps, guzzling glasses of sweet tea. Nana's pretty quiet. Ma said she's always been that way. The one thing I can get her talking about—like most old folks I know—is the past.

"Nana?"

"Yes, noyoner moni?" She uses a term of endearment in our

native language. Nana's the last generation in our family who can speak Bengali fluently. Paw Paw didn't want Ma to learn it because he wanted to focus on bettering his English. And since Ma didn't speak it, I can't either. I've picked up conversational Bengali from Nana and every New Years' I resolve to learn more, but school and life gets in the way.

"Tell me a story, Nana?"

"Which one?"

"The one about the kids."

"Ah," she says, settling back into her chair and staring off into the brown and green jungle that is her yard. "Must have been 2026 . . . or '27 . . . I was young. In my early twenties. I had just moved to America from Bangladesh to escape the floods, not long after marrying your Paw Paw."

I close my eyes to the familiar sound of her lilt. With everything that is going on in the world, this sound is home to me.

"We had already settled in Morgantown. Your Paw Paw had secured an engineering job at a new community solar company. I had begun organizing for climate justice in Bangladesh and wanted to continue that work here. I saw on Twitter that young people had won a big court case in Washington, DC. Julie, Julia, or Juliana . . . some name like that . . . vs. the United States, they called it. These kids— your age, maybe even some younger—had sued the government for their part in allowing climate change to worsen and thus putting their generation's lives in danger. I drove my old Chevrolet Bolt— your grandfather and I had bought it used off a kind neighbor—to Washington, DC, to celebrate. This was a big deal, you know? The American government had avoided getting in trouble for their role in the climate crisis for too long. So I drove to the capital. I parked my car in a garage that probably charged me some ridiculous amount of money that I didn't have to spend," she pauses, immersed in her own memories. "And—I swear—just as I stepped out of the carport, I saw thousands of young people flood the streets. Suddenly there was music, dancing, glitter, and so much laughter everywhere. It was as if we all knew what was coming . . . how big the changes would be . . . how because of the near-century scientists, mothers, organizers, Indigenous peoples had fought for action on climate change, we were able to see some real global momentum."

I smile at this memory. I can imagine the hordes of young people on the streets because I have been one such young person in Washington, DC. Nana passed down her fighting spirit to Ma, who passed it down to me. Nana had much bigger obstacles to fight when she was young, but, as she always tells me, your rights only mean as much as your willingness to defend them.

After that day, a precedent was set for holding the government—and fossil-fuel companies—accountable for the harm they had caused. And every time climate plaintiffs won their cases, they followed the global trend of giving money directly to the impacted communities, including my community in Appalachia.

"I remember that was the day I met Maria . . ."

"Maria Santos, Nana?"

"Mmm. She was so young then. Younger than me even. She had already been in the US at some talks at the UN . . . or maybe the Bank . . ."

"It's so wild that you know her. She's like everyone's hero."

Maria Santos is head of the United Nations. My social studies teacher Ms. Bore is obsessed with her. We've been studying her career for weeks. Maria was a youth climate striker in Manila, back when Nana moved to the US. By 2035, she was the youngest and first LGBTQ+ identifying mayor of Manila. She's such a big deal in part because of how awesome she is, but also because she represents a wave of global government leaders who reimagined ways of measuring success to center communities' needs and well-being . . . especially folks in marginalized communities like the poor part of the city that she grew up in. As a mayor, she impressed people by showing how a leader can stand up for community needs and the good of the planet, and keep everything stable in the city. She served two terms as mayor, then was elected to the National Assembly, then entered the international sphere and moved up from there.

Throughout her career, she has prioritized climate action, especially to meet the demands of the Global South, and emphasized that she is a representative of many people who she listens to to make her decisions. My Nana is one of those people she listens to. I've seen them video calling, both looking seriously at one another and talking about how to meet the needs of the Global South, then suddenly bursting into laughter at some joke or mistake one of them has made. I don't like to tell people that Nana knows her. I like how genuine their friendship is, seemingly untainted by Maria's fame.

"Hero . . . such a silly phrase. You know she'd hate that," Nana says. Both Nana and Maria are anti-idolatry. They say it's why the world got into such a bad place. "You know she just sent a video to my Cele-Frames of her in a night market in Thailand. If only I could find those damn things . . . always losing them . . . and they say they're convenient because they let people show you things right into your glasses . . . well, that's only if you can keep track of the darn things . . . anyway . . . with the hot season ramping up the night markets are as vibrant as ever in the region. Music, food, craftspeople. Local goods and artists have always been treasured in the night markets. Not like here where it took decades for people to wake up to the importance of regional and local trade, and public spaces where people don't have to pay for a right to be there for God's sake . . ." Nana's voice trails off to a mumble, as it usually does when she gets started on the differences between ingenuity in the Global South and laziness in America.

Things are different for me. I can't remember a time when most goods weren't sourced locally. Nana and Ma said there was a time when you had to pay someone money to justify your presence outside of your home. But now it's as if there are more public spaces than private ones. Libraries, parks, community centers. Ma is an artist, on top of being an architect, and one of my favorite paintings of hers is of a man trying to sleep on an uncomfortable-looking bench with a bunch of police officers staring down at him. I like it because it tells a story about the past. It shows me how cities used to be built in ways that made it impossible to be poor. It helps me imagine what they felt like for people, so that I can fight for the better world we have now.

Nana, with her dark brown skin, straight black hair, and Bengali accent was disgusted by this when she arrived in America. She has many stories of fighting for the rights of poor people, Black people, and Indigenous people in her youth. Even as an overwhelmed PhD student, studying Bengali solutions to climate disasters as they consumed her homeland, she devoted her off-hours to strategizing and demonstrating against systemic injustice and for reparations. When she talks to me about those times, she emphasizes most how hard it was to imagine beyond white supremacy and wealth imbalances. She has passed down dozens of Afro- and Indigenous-futurist books to me that helped her imagine a world beyond the injustice she saw. Her copies of these books are so well loved that I have to treat them with extreme care otherwise I fear they'll fall apart in my hands. I lined them up in rainbow order on a shelf next to my bed. Sometimes I lay down on my side and stare at them, thinking of the courage it must have taken the writers to write themselves into our future when powerful forces were trying to invisibilize them.

I look up at Nana. Her eyes are closed, but she is still rocking gently in her chair. I wonder if she is picturing moments from the past. I wonder which moments are coming to her.

I feel hungry for a life of adventure like hers. A life of defending the sacred: home, community, and life.

The sun is slowly sinking behind the Appalachian Mountains. The steady buzz of summer in the South fills the silence. Fireflies flicker in and out of the tall grass around Nana's yard. Nana has lived here in the mountains of West Virginia for fifty years. So much has changed in the world—thanks to folks like Nana who rose above people's tendency to only care about themselves and instead fought for the good of all of us.

I rest my chin on my knees, stare up at the mountains, and exhale.

A Love Letter from
the Clean Energy Future

Mary Anne Hitt

My friends,

It takes my breath away to write these words, but we did it. Rooted in our deep love for this planet and one another, we stepped back from the cliff of irreversible climate change. Families around the globe, including mine and yours, no longer face the specter of fleeing their homes because of ever-worsening climate-driven disasters. The fossil-fuel industry no longer controls the levers of power to corrupt democracy. And we're building a world where everyone has clean air and clean water and access to nature.

We've reached the year 2030, a date that has loomed large for humanity as a threshold that would make or break our climate. As I look back over this pivotal decade in human history, it's clear that, against all odds, the work we did together planted the seeds for the just and sustainable energy future that is now flowering.

A decade ago, we knew if we did this energy transition right, we could prevent runaway climate change, create millions of jobs, and rectify the harms of decades of environmental injustice in communities of color. We also knew we had our work cut out for us—the coal, oil, and gas industries were not going to give up their power or their market share lightly.

As we rolled up our sleeves to prevent a climate emergency, our solutions prioritized investments in communities most harmed by fossil fuels and pollution and those long excluded from economic opportunity. When we won policy solutions that moved us forward but fell short of that vision, we regrouped and tried again. We needed to build so much clean energy infrastructure to avoid a climate apocalypse, and we didn't just build it; we built it with family-sustaining jobs and with an eye toward restitution and reparations. Thanks to you, our kids will be raising their sons and daughters in vibrant, resilient communities full of opportunity. This is how we arrived here:

First, we powered the country with 100 percent clean energy. An electric grid powered by clean energy was the foundation for turning the corner on climate, and the dirty power plants that were the worst contributors to environmental injustice were the first to go. Responding to over a decade of grassroots advocacy, upon entering office President Biden pledged to power the nation with 100 percent clean energy by 2035 (and 80 percent by 2030). In 2022, Congress passed a historic climate bill that set the stage to deliver on that commitment. The legislation was imperfect, a political compromise that included concessions to the fossil-fuel industry that advocates continued to challenge and defeat in the years that followed. But it was also an engine of transformation that provided incentives to retire coal plants, build and deploy renewable energy on an unprecedented scale, manufacture those technologies in the US, and locate new clean energy jobs and projects in low-income communities and areas with economic ties to fossil fuels. The legislation, combined with EPA regulations to tackle harmful air and water pollution from coal and gas plants, put us on our way to phasing out coal and gas by 2035 while ensuring that vulnerable communities experienced the benefits of the transition.

Big states, including California and New York, then set even more aggressive goals, making it clear that a clean energy transition at speed and scale was possible. And since decisions about how we produce electricity are largely made by states, we continued our fifty-state energy-transformation push for a decade.

We finally harnessed the power of offshore wind along the Atlantic coast and solar across the Southeast and Southwest, while scaling up new energy-storage technologies to make clean energy available when it's needed most. Altogether, we made a quantum leap in the scale and scope of the energy transition, produced millions of jobs, and sparked the creation of thousands of new businesses.

Across the Atlantic, Russia's invasion of Ukraine was a wake-up call to Europe and the rest of the world about the perils of relying on fossil fuels from unfriendly nations. Renewable energy development surged as the only solution to provide true, affordable energy security in a time of crisis and beyond.

These sea changes in the clean energy landscape in the US and Europe in 2021 and 2022 began to gather momentum as we rolled toward the middle of the decade. The political, economic, and technological tailwinds had, at last, shifted in our direction. We reached a tipping point, and we had to seize that momentum. We did.

Second, we got well on our way toward electrifying everything. Here in 2030, one of the best parts of the energy transition is that it has made our lives healthier. After social media icons and home improvement TV shows spread the word about how gas stoves create indoor air pollution linked to higher rates of asthma in kids, families rushed to their local home-improvement stores to replace gas ranges with electric induction stovetops. Local governments passed thousands of ordinances calling for all-electric construction in new buildings, states followed suit, and then the EPA issued national standards to phase out gas appliances. New businesses started popping up to help homeowners save money while pulling polluting gas appliances out of their homes. And we created programs at the federal and state level to ensure low-income families could make the switch affordably.

Meanwhile, on the transportation front, states such as California and New Jersey set a 2035 target date for phasing out internal-combustion-engine cars, and national standards followed. States and ultimately the EPA also put in place standards requiring that buses and large trucks go all-electric, which dramatically reduced air pollution in communities of color and big port and shipping centers, including California's Inland Empire, New York City, Chicago, and Los Angeles.

Congress and states also began providing more robust funding for clean and affordable public transit, biking, and walking options. Fleets of electric school buses, postal vehicles, and delivery trucks made communities healthier for kids and families. The number of family-sustaining jobs skyrocketed as Americans were put to work building electric cars, trucks, and buses as well as transit and charging-station infrastructure.

We even began working in earnest to electrify heavy industry, from aluminum to steel to cement. As we transformed those industries, we sparked innovation, revitalized US manufacturing, and began to unwind the long legacy of sickness, environmental injustice, and disinvestment faced by communities living on the fence line of our nation's biggest polluters.

Third, we stopped attempts to expand drilling while we reclaimed abandoned wells, mines, and drilling sites. The oil and gas industry was in a precarious place as we entered the last decade. It was struggling to compete with renewable energy, facing the wrath of communities angry about drilling and pipelines, and grappling with dwindling returns from fracking, which made the industry's finances look more like a pyramid scheme.

Through on-the-ground organizing, we prevented the fossil fuel industry's last-gasp attempt to establish new markets for its products. Even though new oil and gas drilling was included in the federal climate bill, it largely never came to pass as demand for oil and gas continued to fall and grassroots opposition continued. We blocked the construction of more than twenty proposed fracked-gas export terminals in the Gulf South and halted the creation of a new "Cancer Alley" of chemical and plastics plants in the Ohio River valley. We forced the industry to stop drilling next to homes, schools, and communities. And we secured protection from drilling on Indigenous lands, including the Arctic National Wildlife Refuge and Bears Ears National Monument.

Meanwhile, we created jobs for thousands of oil, gas, and coal workers. We put over one hundred thousand people to work plugging millions of abandoned oil and gas wells and addressing methane leaks that were roasting our planet. Congress also directed funds toward reclamation projects and community-led economic development in Appalachia and other coal regions.

Finally, we engaged millions of people in the work for climate justice. Let's be clear: none of this was easy. As we sit here in 2030, the clean and just energy future that we've built together has been the result of millions of people stepping up in their own states and communities.

Indeed, the journey to this moment started long before the past decade. We stand on the shoulders of giants. I was honored to work alongside many of them when, in the first decade of the 2000s, grassroots leaders around the nation stopped a wave of two hundred new coal-fired power plants that were proposed under the auspices of Vice President Cheney's energy task force. Had they been built, that massive coal investment would have locked our nation into another generation of coal-fired electricity and slammed the door shut on the renewable energy revolution that later transformed the twenty-first century.

Momentum grew in the 2010s as a thousand flowers bloomed. Those same grassroots leaders secured retirement of two-thirds of the nation's 530 existing coal plants. Advocates stopped new oil and gas pipelines, won fracking bans, and beat back attempts to drill for oil and gas in iconic and sacred landscapes. Campaigners forced banks and insurance companies to stop propping up the fossil-fuel industry. Community leaders raised the alarm about the health dangers posed by gas appliances in their homes, and the petrochemical and industrial polluters on their fence lines.

Engineers and entrepreneurs planted the seeds of an electric transportation revolution and created renewable energy and home electrification solutions that were cheaper than fossil fuels. Young people took to the streets and to the halls of power to hold older generations accountable and demand action. Climate change became a potent election issue, and the fossil-fuel industry began to lose its legitimacy and acceptability in the public eye.

These are among the many foundation stones of this new world we have built, and the work continues. I know all this seemed impossible a decade ago, when it felt as if everything was falling apart and our climate might be doomed. But everything we did mattered. All of it.

We now know that we're going to keep global temperature rise below the most dangerous tipping points that climate scientists warned us about a decade ago. We can look our kids in the eye and tell them that we didn't let them down. Now we can watch their dreams unfold.

As all our great spiritual traditions have taught us, new beginnings are often born during our most difficult days. We created something beautiful out of those hard years a decade ago, when we were staring into the climate abyss. We stepped back from the edge, determined to write a new and better chapter in our shared future. Of course, we have more work to do. But we're doing that work from a foundation we built together. I can't wait to see what we'll do next.

Different Ways of Measuring

ON RENUNCIATION AND ABUNDANCE

Rebecca Solnit and Thelma Young Lutunatabua

Most people believe a clean-energy future will require everyone to make do with less, but it actually means we can have better things. There are obviously a lot of barriers to accomplishing this plan. I tell people what is technically necessary, and they tell me about political barriers. As naive or implausible as it sounds, we have to figure out how to remove all of those barriers—one at a time, and then, hopefully, many at once. Policymakers have to change what they believe is possible in the current political and economic climate. If what is politically possible is the extent of their ambition, everyone is doomed.

—**Saul Griffith,** *Electrify*

> It is some very effective marketing that has con-
> vinced so many of us that getting off of fossil fu-
> els is a sacrifice as opposed to a money-saving,
> peace-promoting, water-protecting, health-im-
> proving, technological leap forward.
>
> —Dr. Elizabeth Sawin

Rebecca: What if the climate crisis requires us to give up the things we don't love and the things that make us poorer, not richer? What if we have to give up the foul contamination around fossil-fuel extraction, the heavy metals people inhale when coal is burned around them, the oil refineries that contaminate the communities of color around them from the Gulf of Mexico to California? What if the people of Richmond in my own home region, the Bay Area, didn't have emergency alerts where they were supposed to seal their homes because of a refinery leak? What if the incidence of asthma in kids went way down, and we stopped losing almost nine million people a year to air pollution worldwide? What if that moment when the pandemic shut down so much fossil-fuel burning that people in parts of northern India saw the Himalayas for the first time in decades became permanent?

Thelma: It's so important to be able to differentiate between what is necessary to hold in abundance and what is okay to let go. There are treasures we need to celebrate in their plentifulness—coral reefs, bird song, block parties, and more. And then there are the items, beliefs, symbols that it is okay to allow to diminish. I love Berlin in the summertime and going on long walks past places that have minimal reminders of where the wall used to rip through, but now there are community gardens, children's daycare centers, ice cream shops, and more. When the old world is broken down, such wonders can sprout up.

Those who have made their vast fortunes through the consumption-driven frenzy of the fossil-fuel age want us to be scared about having less. The Instagram-fueled ideal life is just a mirage to begin with that is only available to a fraction of humanity. Consumption disconnects us from community. When you're sick, you order soup online instead of a friend bringing it over to you. If any good buds of truth came out of the pandemic and perpetual climate disasters, it's the reminder that communities and healthy ecosystems are how we survive. Especially for many people who didn't experience the social safety net of wealthy governments or the privileged convenience of remote work—they made it through the dark times of the past few years through neighborhood exchanges, new gardens, fishing poles, and other networks of mutual aid. The predominant Western world order wants us to all put up walls around ourselves—but such glorious things can happen when they come tumbling down.

How do we shift societal perceptions of what a "good life" is? What's holding us back—is it corporations and governments or is it our own blinders?

Rebecca: Many years ago, my wonderful Utah-environmentalist friend Chip Ward used the phrase "the tyranny of the quantifiable," and I've borrowed it and worked it hard ever since. What you say, Thelma, is that we are told that we live in an age of affluence and abundance and we should fear change. There is plenty of quantifiable abundance out there—as stuff, disposable stuff, fast fashion, cheap furniture, rapid travel, packaging and litter and garbage heaps and the plastic patch in the Pacific. The carbon dioxide we've added to the upper atmosphere is in its own way a kind of garbage dump, a residue of our fossil-fuel consumption.

But this age is poor in so many ways—there is so much lone-liness and disconnection, from love, friendship, community, from the natural world, from moral and visual beauty, so much hopeless-ness, so many who don't have enough in a world full of people who have too much. What if that is what we need to renounce? What if the climate crisis requires renouncing not this version of wealth but its underlying poverty? Overconsumption requires sweatshops in the Global South, extraction of resources in ways that devastate local communities, and other forms of impoverishment of both the human and nonhuman, of the bodies and lives there and the prin-ciples and imaginations here. What if we measured our wealth in other ways, as confidence in the future, as the clarity of the air and its breathability, as pride in one's community and country, as integ-rity in our material and moral lives, as having seen the summer's shooting stars or throwing moon-viewing parties or watching the migratory birds arrive, as friendship and the sense of safety that means help is there when needed, as honor and dignity and a mean-ingful life?

In your terms, what if the walls came down?

Thelma: I agree, we need different ways of measuring. One of our other contributors, adrienne maree brown, speaks often about how we need to be better at learning to be satisfied. When do we say, *Enough is enough*, and *I feel full*? A dear Fijian friend, the incompa-rable Alisi Rabukawaqa Nacewa, has told the story of when she first learned from her grandmother the importance of only taking what you need from the local reef. The principles were grounded in care of nature and care of community. Other people would also need the seafood and the creatures need to be able to reproduce as well, so don't overfill your bucket just because you can. Some people might want to label this as "austerity"—but it's really so much deeper—it comes down to principles of how we see ourselves in relationship to other people and other species. This is what you're speaking to as well, if we strengthen our communities and ecosystems—what abundance can that bring? This is the direction I hope we go as we consider climate policies on an individual, national, and global level—how are our actions impacting others? Am I taking more than I need? Many industrial countries right now are continuing to open new fossil-fuel infrastructure, even though international agencies, vulnerable countries, and economists tell them it makes no sense. Limitation does not equal deprivation, and I wish the Western world learned to embrace this.

Rebecca: How can you ever have enough if you can slip and fall into having nothing? In a society where no one could fall that far, could we all feel that we had enough, and we could focus more on living up to our ideals rather than piling up affluence as a wall against that fear? Affluence also becomes a means of hiding from the consequences of inequality; it is all about building walls.

I go back to how we've learned to measure the wrong things— how much we have quantified security as personal wealth and its ability to isolate rather than confidence in the collective future and the ability to connect. We also apply the lessons of money and material objects to the things of the spirit—it is true that if I take your belongings I have more of those things, that if I give away my own possessions or money I have less. But to give more love, more praise, more joy, does not deplete me, and to give creates the mutuality that means these things will be given back to me.

And if climate teaches one thing it's that everything is connected. The fossil fuel we burn in the industrialized world changes temperatures in the Arctic and the tropics, melts ice at the poles that impacts sea level rise in the Pacific islands. Climate chaos is a catastrophe but also a teacher, and its first lesson is that everything is connected, which is both the beautiful dream of mutuality and the nightmare of runaway consequences. We need to learn to see that interconnectedness, and as we both believe so deeply, this is why this crisis, like every crisis, is in part a storytelling crisis.

Thelma: Truly, and for far too long the rhetoric in many climate spaces is we need less—"Drive less, eat less, turn off our lights." It's this language that implies to people that in order to care about the climate you must have a lower quality of life. When really, the calls to action need to be questions of connection: Are you getting to know your neighbor? If a flood came through tomorrow, would you be ready to assist each other? Are you holding politicians accountable for how their policies impact frontline communities? Are you getting to know the various names and patterns of the plants in your area?

If we can shift our task list toward mutuality, we will start feeling the spaciousness of the journey ahead. Once we shift our worldview to center relationships, we will also better be able to grasp what needs to be nurtured to grow and what can be left behind. Through tackling the climate crisis, I hope it will lead us down eye-opening paths of abundance. Let's leave behind what created the monster— white supremacy, dangerous extractivism, patriarchy, and more. Hopefully, the next generations will hear more elaborate songs from nature, know better ways to repair and share, and find a more sumptuous and intertwined way of being.

TAKE THIS
WITH YOU

It is possible that the next Buddha will not take the form of an individual. The next Buddha may take the form of a community—a community practicing understanding and loving kindness, a community practicing mindful living. This may be the most important thing we can do for the survival of the Earth.

—Thich Nhat Hanh

We are beginning to understand that the world is always being made fresh and never finished; that activism can be the journey rather than the arrival; that struggle doesn't always have to be confrontational but can take the form of reaching out to find common ground with the many others in our society who are also seeking ways out from alienation, isolation, privatization, and dehumanization by corporate globalization.

—Grace Lee Boggs

We don't have to wait for anything at all. What we have to do is start.

—Octavia Butler

Packing (and Unpacking) for an Emergency

Rebecca Solnit

I live in earthquake country, where we're all urged to prepare a kit to evacuate or shelter in place for when "the big one" shakes us up. Not far away, people live where climate chaos–fueled wildfires mean they may have to flee on short notice with what they can grab. In other parts of North America, hurricane evacuation has long been part of life. Mostly, when we talk about packing for an emergency, we talk about tangible goods—flashlights, first-aid kits, water, documents, and keepsakes, things like that.

But there's another way we pack—and unpack—for emergencies. The stories, ideas, values, aspirations, and facts we carry strengthen or weaken us, connect or disconnect us, motivate or demoralize us. They not only prepare us to face emergencies but also to change whether and how they happen. That is, they make us who we are, and who we are has everything to do with who and what survives.

What will you take with you in an emergency? What are you packing for the emergency that is the age of climate change? A crisis tests stories and beliefs. Some break, some get stronger, some change. Some stories are life rafts or desert springs; some are poison or prison or heavier than we can carry where we need to go.

We have tried to pack this book like a kit for you to carry with you, and we filled it with voices offering views and visions we thought were brilliant or lifesaving or eye-opening, ideas and facts about the reality of the situation and what we can do about it, examples of people doing it, invitations to power and possibility.

We wanted to address what sometimes gets called climate grief, but we believe that you can't address the grief without addressing the climate. That is, if you're devastated because your house is on fire, finding ways to put that fire out or rescue the people who are trapped inside is relevant. If you're anxious because you fear your house is going to burn down, finding people committed to fire prevention or building less flammable houses matters.

One underlying story worth unpacking tells us we're too fragile to face hard things, that if we are sad we will break, and we should never be sad because it's a kind of failure. But sadness is an appropriate response to threat and damage to what you love, and the underlying love matters. That's why this book includes Roshi Joan Halifax's reminder that grief and fear "can be a kind of doorway" and voices from resolute people on the front lines of the climate crisis. Many in this book spoke about grief that did not stop them but made them more committed.

In now famous words, prison abolitionist Mariame Kaba tells us that "hope is a discipline." As I noted in the introduction, this isn't hope as optimism which assumes everything will be fine and nothing is required of us. That's only the flip side of pessimism and despair, which likewise require nothing of us and buffer us from uncertainty. The future is a night in which we cannot see far; we can only navigate it by looking to the past, where we can count our victories, learn from our defeats, measure change, and see how power grows and imagination shifts. This is as true of the climate movement as of anything; the movement has grown, achieved extraordinary things, improved the possibilities, awoken the world. It still has far to go, but these accomplishments should fortify our confidence to continue.

There will be losses; there are losses behind us and ahead of us; but there are nevertheless things worth protecting and will be every step of the way. Julian Aguon asserts, "Part of our work as people who dare to believe we can save the world is to prepare our wills to withstand some losing so that we may lose and set out again, anyhow." One of the burdens no one needs to carry is perfectionism, including the idea that if we can't save everything we can't save anything, that if we don't win perfect, comprehensive, final victories, then we must be losing.

Most victories are partial or compromised, nearly all of them are interim, because the story continues. Often, winning is made up of what did not happen and thus what cannot be seen, and we only know about it through story. The radical poet Muriel Rukeyser said long ago, "The universe is made of stories, not of atoms." Choosing the stories you, yourself, are made of is crucial work as we enter this unfinished story of how human beings responded to the greatest emergency our species has ever faced.

Mary Annaïse Heglar, whose tenacity is matched only by her brilliance, declared in 2022, "If you're worried that it's too late to do anything about climate change and we should all just give up, I have great news for you: that day is not coming in your lifetime. As long as you have breath in your body, you will have work to do." Mary lives in New Orleans, which has a lot to teach about community in crisis. In 2005, Hurricane Katrina broke the levees that protect the city, and 80 percent of this city surrounded by lake, river, canal, and wetland went under water. The authorities largely failed the stranded, who were mostly poor people who couldn't afford to evacuate.

But locals and even people from as far away as Texas got their boats and got into those waters to see whom they could rescue from the rooftops and the freeway overpasses. No one thought they could save everyone; everyone thought it was worth trying to save someone; a lot of people lived and got to safety because strangers plunged into those murky waters to see what they could do. In any crisis, we have to plunge in, and only while doing so will we find what we can save. We can't aid anything or anyone unless we do.

"We" is a crucial word. Two intertwined themes run through all the pieces in this book. One is love, and the other is community. Individualism can seem like an invitation to travel light, but it often keeps you from getting anywhere at all. It strands you on your own. One response to climate says all we can take care of is our own feelings or ourselves, but it's the connections—to places, people, movements, ideals—that fortify us to face the situation and make it possible to change that situation.

Corporate-backed media would like us to think that all we can do is lead low-impact, modest lives—watch our climate footprint, recycle our packaging. They'd like us to stay disconnected, so we have no power to confront them and their misinformation and damage. That version leaves out the power of voting, organizing, campaigning, joining, the power we have together, and the joy in that power and connection.

Bill McKibben has long said, when people ask him what's the most significant thing you can do for the climate, "Stop being an individual; join something." There are deeper layers inside individualism and its alternatives. As Fenton Lutunatabua of the Pacific Climate Warriors reminds us in one of this book's conversations, "So much of this story about individualism needs to be left behind. The future needs to be one that's collective and communal."

There's a Buddhist anecdote I love, in which the Buddha's cousin and student Ananda remarks to him that friendships and camaraderie with admirable people are half the holy life. The Buddha replies that no, they're the whole thing. A lot of climate grief seems to be connected to loneliness, to the sense that no one cares, or the people in power don't care. Being with people who do care helps, and so does evidence that the majority of people do and want to see climate action.

But community is crucial in the most practical terms, too. The most essential equipment for surviving a sudden upheaval is the neighborhood, civil society, the people who will dig you out of the wreckage or start the soup kitchen with you or help rebuild the town or check on you in the heat wave or the blizzard. This means that some of the most pleasant things imaginable—building relationships and friendships, getting to know the people around you—are also the most crucial. In societies obsessed with productivity, these get classified as leisure or worse—a waste of time, a distraction, frivolity—or they get called social capital, as though you could pile it up in a bank somewhere. But they are how we build and tend that thing called society, and, in the long emergency that is human life in this era of climate change, that's crucial.

The climate movement and climate activism are about relationships and alliances; they, too, are a source of community. I have been to plenty of tedious meetings and protests that didn't feel like they accomplished anything I could see, and I've seen my share of infighting. But I've also been at moments when I could feel the crowd come together with the mysterious alignment that made it a powerful, united force and the day a day of history. I've seen the world get changed by the power of organizing.

I've marched with and listened to and met some of the most heroic, most generous, most brilliant, most principled people I could ever hope to encounter. I take inspiration from the climate organizers, scientists, policy experts, and journalists around the world I watch, read, and listen to from afar. The long view has helped, including watching the growth of a global climate movement, seeing the building of power, insight, alliances, and visions; seeing the consequences of that in policy and practice; and seeing the dedication of so many people manifest in so many ways.

With the information we take in—whom we read, whom we listen to—we build a kind of virtual community. This is where society becomes story, and choosing your sources and stories is crucial to your well-being and participation. There has been a lot of misinformation from all directions about climate, and that misinformation can depress, paralyze, confuse, and otherwise harm. We had decades of climate denial from the beneficiaries of climate destruction and the status quo from all directions—the fossil-fuel industry, related industries, financial titans, and the politicians who took their money.

Now, they pretend to care and offer responses that usually involve delaying what's necessary, or they promote false and inadequate solutions. *Trust us*, they say, *do nothing; we will make the decisions and choose the solutions.* There is no substitute for a radical reduction of fossil-fuel consumption, and therefore of extraction, as the heart of any good climate policy, but you wouldn't know that if you heard all their schemes to continue the drilling and pumping and burning. And it's the power of ordinary people that will bring about this transformation.

At the other end of the spectrum are stories that it's too late, that it's all over, that there's nothing left to do. Misinterpretations of the data lead some to declare that the Earth itself is going to die or human beings are going to die out in the near future, stories that clash with what the evidence tells us. They make people feel terrible, and they make them passive. Doomism, as climate scientist Michael Mann calls it, becomes a self-fulfilling prophecy if it prevents the action that can shift us away from the worst-case scenarios. Despair is the opposite of complacency, but both can lead to inaction. If you don't believe there's a fire or if you don't believe the fire can be put out, you don't join the firefighters.

We are all called upon to balance a sense of danger with a sense of possibility, to pack some of both in our kits and not let one outweigh the other. Too much of either leads to inaction. To do that means recognizing we are capable of complexity, of holding many emotions, of understanding the sheer uncertainty of the future and the possibility of participating in shaping it, of knowing that since we cannot know outcomes ahead of time, we are also making decisions about who to be and how to live.

Stories about how others before you, individuals, communities, cultures, struggled and survived against the odds, stories of victories on the issues you care about, of commitment and integrity, can give you wings or be your compass. I know a lot of us try to cope with overwhelming political realities by trying to master all the facts, and that can turn into doomscrolling or just overload. With climate, the issue—as science, as politics, as technology, as culture—is more than anyone can grasp. I limit my diet of bad news, because I don't need to know every bit of it, and too much of it is undermining, not motivating.

Finally, this book is a love story. Every single person in this book does their work researching or educating or organizing or storytelling because of love. Love for the coming generations, love for the community around them, love for the natural world, love for justice, for truth, for possibility. Underneath the fury at fossil-fuel companies: love. Behind the commitment to organize for year after year: love. All through the demanding research scientists do: love. Find what you love and make sure you never lose sight of it. It's the well that you drink from, the touchstone that reminds you who you are and why you're here.

We are sometimes told that pleasure, beauty, and joy are not necessary, but they help us keep our bearings, keep our strength, keep our vision. And there can be all these things in the future we're making. The climate movement worked for years to get people to see the current and future damage we faced, to emphasize danger and loss. We've succeeded so well that I believe we now need to balance that with attention to the places that are still thriving, still beautiful, still provide respite and a sense of order.

The world will not stop being beautiful, not stop having sunrises and full moons, light pouring through clouds; and while some places may be scarred or drowned and for a time lifeless, in others, seeds will break through soil, flowers will bloom, birds will soar, new life will be born, babies will learn to walk.

We started calling it climate chaos because it's breaking up the patterns of life, but not all of them everywhere. The climate crisis, to give it another name, will not end suddenly, and equipping yourself to keep going means finding what feeds your heart and soul. "Even a wounded world is feeding us," writes the Indigenous plant scientist Robin Wall Kimmerer. "Even a wounded world holds us, giving us moments of wonder and joy. I choose joy over despair. Not because I have my head in the sand, but because joy is what the Earth gives me daily and I must return the gift."

Not Only a Danger but a Promise

Thelma Young Lutunatabua

To be pregnant is to be acquainted with death. Through the process of creation, one has to also be ready for the realities of destruction. It's estimated that one-third of all pregnancies end in miscarriage, with our own bodies declaring that now is not the time for life. There are also all the health dangers that come with pregnancy, including even the horrifying fact that a woman is more likely to be murdered while pregnant. In the first trimester of my pregnancy, in Fiji, amid all the joyous expectations, there was also the haunting cloud that told my body to brace for the potential of seeing blood, to be ready for an end. We're also unfortunately in a time when some US states are passing more restrictive laws aimed to control a woman's body. The second half of my pregnancy is in Texas, so there are the hovering fears of *What if something happens, and the patriarchal systems demand a sacrifice? What will happen to my life?*

It is a complex experience to work on a book about climate hope while actively growing my first child. It demands an honest meeting with the realities of the future. Like many parents, there is the tumult of wondering what the world will look like for this upcoming generation. Yet, while also being in a space of creation, I feel a unique sense of what could become. There are many of us in newer generations that recognize that the systems of the past do not work for us, so we are looking for what needs to be seeded instead. There are so many entrenched systems of white supremacy that must be brought down. Those that said it's okay to take and take and take and never return. Those that diminished the sacredness of community and the diversity of life. Those that preached the gospel of sacrifice zones. Something else must bloom.

To be pregnant is also to be acquainted with rebirth. There are books on my shelf about childbirth and raising kids that I haven't opened yet. I'm still spiritually processing that this is a moment of total life transformation. The more I fight it, the more it will eventually swallow me up. I'm reckoning with the fact that many women have not been trained to trust and understand our bodies. We have become disconnected from natural processes and inter-generational wisdom. Yet, despite my ignorance, my body knows exactly what to do. My life is in a period of transformation. Stagnation is undesirable and unavoidable. I must give myself up to the processes underway.

A massive rebirth is needed right now for much of the industrialized world—one that will justly transition us beyond the extractive dominance of fossil fuels and into a new way of being. Over the past few years I've found myself rereading the manifesto that Grace Lee Boggs and other women leaders wrote in 2005 that proclaimed, "Another world is necessary, another world is possible, another world has already started." In it they declare, "This universal crisis is not only a danger but a promise, an opportunity to advance ourselves and our societies to a new level, based on a new vision, new principles and values." With each new pipeline approved, with each new gas permit granted, world leaders show that they are not ready for this rebirth. But the Earth knows what's necessary and is continuing to send shock signals that we cannot keep on postponing this transition. How do we keep on rolling this rebirth along, even despite all that is politically, economically, environmentally against us?

There is a profusion of blockages working to hold us back. To care about the climate crisis is to be in a steady state of managing grief and pain. And this is where my own process of creation has become a teacher. I don't think I fully understood hope until my first miscarriage. The heartbeat was there, and then three hours later I stuck my hand into the hospital toilet to retrieve the clump that had just come out of me, in an adrenaline panic that it held what remained of the child that had been. While lying in a bed at 3:00 a.m., managing the subsequent bleeding and staring at the bleached once-pink bed sheets turned curtains, I kept on thinking of something that Rebecca had once told me, "You don't plant a seed unless you hope it will grow." My husband and I had planted this seed with an abundance of love and hope, yet it did not grow. The following days were a haze of recovery, slow grief, and more blood. Miscarriages are something that are still often so hushed up and that go unspoken, and you don't find out how many people have also experienced this until it happens to you, and you start tenderly, hesitatingly start telling people. I didn't know how people keep going with trying to have a child after experiencing this sort of loss. What was so shocking to me was how quickly my body recovered and got back into its normal cycle. My body and the shift in hormones were telling me it was okay, and that it was ready before my spirit was. Earlier in this book, I interviewed my husband about the Pacific Climate Warriors' motto, *We are not drowning, we are fighting*, and we applied that to our own familial efforts. We chose to not be swallowed up by the waves of present and potential pain.

We honored our emotions, and then each day brought a choice to make to not be swallowed. Sometimes you plant seeds, and they do not grow. But the decision to keep on planting and planting taught me that hope is about the everyday steps taken beyond the dark wells of sorrow. It's not that we avoided or disregarded the grief, but we also did not stop at grief.

With this latest pregnancy, I safely passed the point at which my previous miscarriage had happened during the same week that we reached the end of cyclone season. There was the simultaneous sigh of immense relief that came from both—that destruction had been averted for now.

The future will no doubt hold its fair share of pain. Even if we can transition off fossil fuels right away, it is uncertain what levels of greenhouse gases are locked into the atmosphere, how quickly our ecological systems will right themselves (though some scientific estimates look more promising than others). Yes, there will be more heat waves, more monstrous cyclones, more crop failures. The current status of our warmed planet is already horrifying, and the future projections even more so. Yet we are fools wrapped in privilege if we think that there hasn't always been suffering. I think of my child's ancestors, who endured so many nightmares. I know my child will have to be a fighter, or even better—an organizer. Someone who knows that survival is tied to community and care. Someone who has listened to their grandmother and understands the ways of plants. Someone who has seen coral reefs relentlessly fight their way back and launch new blooms.

Pregnancy is teaching me how to hold destruction and creation simultaneously. This translates to my daily life during our current era of climate crisis—I am heartbroken at the lives destroyed by floods, drought, and greed, yet I also praise the new victories won, the pipelines stopped, the community solar projects built.

I completely honor and respect people's decisions to not have children because of climate fears. For me, having a child in 2022 is an act of radical hope. Cynicism seems, often, to be a by-product of toxic masculinity—the indifferent rebel/loner who turns his back on the world is often uplifted as the epitome of cool. As a mother, cynicism is not an option. My hope is tied to an existing practice of refusing to allow apocalyptic prophecies of the future to come to pass. It is a literal way of having more skin in the game. There is no choice for me but to ensure as many families as possible can safely make it through this rebirth.

ADDENDUM

My son was born two months after we turned in the first draft of this book. Like many great beginnings, his first moments were heralded by a group of competent women who ensured that he and I were safe as well as thriving and respected. Looking back on this essay and book from the realm of postpartum has reinforced the key lessons that surfaced in its creation.

The first is that we can't do this alone. Whether creating new life or whole new socioeconomic systems, teams of people all giving their talents are needed. Since giving birth, I have been surrounded by care, and this is how every emergence should be. The second is that this work has to be grounded in love. That doesn't mean that the pursuit of climate justice will be easy and full of rainbows. Instead, like motherhood, love ensures continuity—that the difficult moments are made bearable and the bright moments nurture and sustain.

· Last, I am learning the importance of surrendering to my new ways of being. Accepting new possibilities can be a real learning curve, and it can broaden with new horizons. Right now, my body is adjusting to the rhythms of feedings every two hours. I can choose to fight that my life has now become this, or I can accept and find joy in growing baby cheeks as well as endurance in the sleepless hours. The world will change in the coming decades, there's no doubt about it. Don't surrender to the disasters and corrupt politicians. No—surrender to the new works of social change already showing us hope and possibility.

Acknowledgments

My first thanks go to Thelma, with whom I struck up a long-distance conversation in mid-2020 when we realized we had shared views about the impact of the climate crisis on emotions and the role that good facts and frameworks can play. We spent more than a year talking before we named our project Not Too Late and launched it as a website and social media in May 2022. This book was planned during an enchanting week when she came to stay with me and we drank tea, snacked our way around San Francisco, and sketched out the climate anthology of our dreams. She has remained throughout a joy as a friend, a collaborator, and a person from whom I've learned much.

My second go to a wonderful gathering of old and new friends in March 2022: when I told them I'd stopped writing books to try to be a better climate activist and told them about Not Too Late, they, in unison, told me turn it into a book, so we did. For that nudge, thank you to Saket Soni of Resilience Force, Annie Leonard of Greenpeace and her daughter Dewi Zarni, and the writers Lauren Markham, Ben Gucciardi, and Arlie and Adam Hochschild.

My third to Anthony Arnove and Haymarket Books, which had the nimbleness to take up the resultant book proposal and work with us to speed this book on its way in a little over six months from first draft to published volume, and to the wondrous Caroline Luft and Rachel Cohen, respectively among the best copyeditors and book designers I've ever worked with.

My fourth to the twenty contributors to this book, who are my friends, heroes, allies, and teachers. We were honored and exhilarated to gather their voices and perspectives. Beyond their voices—originating from Guam to West Virginia, Bangladesh to the Philippines to Louisiana—lies the vast global climate movement, which has grown in size and vision and strategy over the many years we've been part of it. Some were already dear friends, some became new friends; all of them were generous with time and ideas.

Lastly, to dear friends and climate allies who have not infrequently been the same people, and to the climate movement, which I've seen grow in power, sophistication, and vision over the past twenty years, and which is maybe not something that *has* hope so much as it makes hope for us all.

—Rebecca

Collaborating with Rebecca before this project and throughout has been a deep experience in care and possibility. Not only does she approach this work with rigorousness and wisdom, but she also puts equal attention on making sure those around her feel nourished and in safe keeping. I am so deeply grateful she saw me as a potential partner and has treated me as an equal all along the way.

I also want to express gratitude for my family. My husband Fenton (who also agreed to be part of this book) is continually showcasing how to live in a way that honors community. We met through doing climate activism work many years ago, and he has been my partner in justice and life. Thanks also to my mother-in-law Angeline and my parents Douglas and Elaine, who live so selflessly and have cared for me throughout this journey. And, of course, to my son Anders, who has literally been with me through every step of this book and I know will continually teach me about the power of creation and hope.

This book could not exist without the ongoing perseverance of the global climate movement. From Kenya to Samoa to London and more, there is a massive spread of people doing what they can to protect our homes and those we love. It's so important to look beyond the usual images of the climate activist and to behold the full scope of those in this struggle—artists, designers, local officials trying to move good policies ahead, mutual aid groups, community gardeners, disaster relief workers, and so many more. We're more powerful than we realize. There is an infinite number of talents out there, and so an infinite number of ways people are pushing against destruction. Thank you to every climate activist who has shared a meal with me, marched with me, shared your story, and especially laughed with me.

And a quick thanks to the Seguin Public Library and KEXP for being there as I pushed through the tough moments putting together this book.

—Thelma

Epigraph Sources

JOIN US

Tarana Burke in conversation with Ai-Jen Poo, "The Future of Hope 5," April 21, 2022, interview by Krista Tippett, *On Being*, podcast, https://onbeing.org/programs/ai-jen-poo-and-tarana-burke-the-future-of-hope-5.

Pirkei Avot, cited by Rabbi Jill Jacobs in "Pirkei Avot: Ethics of Our Fathers," My Jewish Learning, myjewishlearning.com/article/pirkei-avot-ethics-of-our-fathers.

Jia Tolentino, interview by Christopher Bollen, *Interview* magazine, July 8, 2020.

WE HAVE THE SOLUTIONS

Donella Meadows, "Is the Future Our Choice or Our Fate?," April 19, 2001, https://donellameadows.org/archives/is-the-future-our-choice-or-our-fate.

David Graeber, *The Utopia of Rules: On Technology, Stupidity, and the Secret Joys of Bureaucracy* (New York: Melville House, 2015), 89.

James Baldwin, "As Much Truth As One Can Bear," *New York Times Book Review*, January 14, 1962.

Mike Davis, quoted in Dana Goodyear, "Mike Davis in the Age of Catastrophe," *New Yorker*, April 4, 2020.

FRAMEWORKS OF POSSIBILITY

Bayo Akomolafe, Facebook post, March 3, 2022.

Ocean Vuong, "A Love Letter for Our Community," Thich Nhat Hanh Foundation, February 17, 2022, https://thichnhathanhfoundation.org/blog/2022/2/17/a-love-letter-for-our-community-by-ocean-vuong.

Reverend angel Kyodo williams, Facebook post, November 16, 2016.

Rachel Cargle, in conversation with Leah Thomas, Atmos, May 7, 2021, https://atmos.earth/rachel-cargle-leah-thomas-intersectional-environmentalist.

THE FUTURE WE WANT

Angela Davis, quoting her mother's words to her when she was young, May 24, 2022, San Francisco, CA.

Robin Wall Kimmerer, "Ancient Green: Moss, Climate, and Deep Time," Emergence Magazine, April 20, 2022, https://emergence-magazine.org/essay/ancient-green.

Ursula K. Le Guin, speech in acceptance of the National Book Foundation Medal for Distinguished Contribution to American Letters, November 19, 2014, San Francisco, CA.

Howard Zinn, "The Optimism of Uncertainty" (1988), in *Failure to Quit: Reflections of an Optimistic Historian*, 2nd ed. (Chicago: Haymarket Books, 2014).

Saul Griffith, *Electrify: An Optimist's Playbook for Our Clean-Energy Future* (Cambridge, MA: MIT Press, 2022).

Dr. Elizabeth Sawin, Twitter, April 9, 2022, https://twitter.com/beth-sawin/status/1512964333151653889.

TAKE THIS WITH YOU

Thich Nhat Hanh, "The Next Buddha May Be a Sangha," speech, Day of Mindfulness, October 1993, Woodacre, California.

Grace Lee Boggs, *The Next Great American Revolution: Sustainable Activism for the Twenty-First Century* (Oakland: University of California Press, 2012).

Octavia Butler, cited in Lynell George, *A Handful of Earth, a Handful of Sky: The World of Octavia E. Butler* (Los Angeles: Angel City Press, 2022).

Resources

Visit haymarketbooks.org/books/2000-not-too-late to download a free study guide to this book, or visit nottoolateclimate.com, where you will also find our guide to additional sources of news, information and skills, literature and art, and action organizations.

Contributor Biographies

Julian Aguon is an Indigenous (Chamorro) human rights lawyer and writer from Guam. He is the founder of Blue Ocean Law, a progressive firm that works at the intersection of Indigenous rights and environmental justice. He serves on the global advisory council of Progressive International. He was a finalist for the 2022 Pulitzer Prize for Commentary.

Jade Begay, Diné and Tesuque Pueblo, is an Indigenous rights and climate policy expert, organizer, filmmaker, and the climate justice campaign director at NDN Collective. She has partnered with Tribal Nations and Indigenous communities on issues like climate justice and Indigenous self-determination. In 2021, she was appointed by President Biden to serve on the first White House Environmental Justice Advisory Council.

adrienne maree brown grows healing ideas in public through her multigenre writing, her music, and her podcasts. Informed by twenty-five years of movement facilitation, somatics, Octavia E. Butler scholarship, and her work as a doula, adrienne has nurtured Emergent Strategy, Pleasure Activism, Radical Imagination, and Transformative Justice as ideas and practices for transformation. She is the author/editor of seven published texts and the founder

205

of the Emergent Strategy Ideation Institute, where she is now the writer in residence.

Edward R. Carr is a geographer and anthropologist connecting academia, policy, and implementation to address the climate crisis. Based at Clark University, he was a lead author of the Working Group II contribution to the IPCC Sixth Assessment Report and advises the Global Environment Facility on its Climate Change Adaptation portfolio.

Renato Redentor Constantino has led networks, campaigns, and organizations on international climate policy for three decades. He contributed to the anthologies *Harvest Moon: Poems and Stories from the Edge of the Climate Crisis* (2022); *Letters to the Earth* (2019); *Humanity* (2018); *Agam: Filipino Narratives on Uncertainty and Climate Change* (2014); and *The World Can Be Changed: An Anthology for Posterity* (2004). He is the author of *The Poverty of Memory: Essays on History and Empire* (2006).

Joëlle Gergis, PhD, is an award-winning climate scientist and author, based at the Australian National University. She served as a lead author on the United Nations' Intergovernmental Panel on the Climate Change's Sixth Assessment Report. Her latest book is *Humanity's Moment: A Climate Scientist's Case for Hope.*

Jacquelyn Gill, PhD, is an associate professor at the University of Maine's Climate Change Institute, where she researches and teaches about climate change, extinction, and biodiversity. She co-founded the March for Science and is an award-winning science communicator, podcaster, essayist, and environmental advocate. She is definitely on #TeamMuskOx.

Roshi Joan Halifax, PhD, is a Buddhist teacher, social activist, author, founder and head teacher of Upaya Zen Center in Santa Fe, New Mexico, and in her early years was an anthropologist. She is director of the Project on Being with Dying and founder of the Upaya Prison Project, which develops programs on meditation for prisoners. Her books include *The Fruitful Darkness: A Journey through Buddhist Practice*; *Being with Dying: Cultivating Compassion and Wisdom in the Presence of Death*; *Standing at the Edge: Finding Freedom Where Fear and Courage Meet*; *Sophie Learns to Be Brave*.

Mary Annaïse Heglar is a climate justice writer and the cohost and cocreator of the Hot Take podcast and newsletter. Her essays have appeared in *Rolling Stone, Boston Globe, WIRED, New Republic,* and other outlets. She was the inaugural writer in residence at the Earth Institute at Columbia University. She has served as an adjunct professor at Columbia University and will take on an adjunct position at Tulane University. She is based in New Orleans, Louisiana.

Mary Anne Hitt is a climate strategist and writer. She is the senior director of Climate Imperative. She has over twenty-five years of experience building and leading effective campaigns and organizations, including serving at Sierra Club as national director of campaigns and director of the Beyond Coal Campaign. She lives in West Virginia.

Nikayla Jefferson was born and raised in San Diego and, after undergraduate at UC Santa Barbara, cofounded the Sunrise San Diego hub. She has spent the last few years bouncing between graduate school at UC Santa Barbara, couches and spare bedrooms, the American highway system, and Sunrise Movement staff. She is a writer with an aim to peak at age fifty-five.

Kathy Jeñtil-Kijiner is a poet from the Republic of the Marshall Islands. She also serves as a climate envoy for the Ministry of the Environment and is the cofounder and current director of the nonprofit Jo-Jikum, a youth environmental nonprofit.

Antonia Juhasz is a leading energy and climate author and investigative journalist. An award-winning writer, her bylines include *Rolling Stone*, *Harper's*, *Newsweek*, *New York Times*, *Los Angeles Times*, *The Atlantic*, *CNN.com*, *The Nation*, *Ms.*, *The Advocate*, *The Guardian*, and many more. Antonia is the author of three books: *Black Tide* (2011); *The Tyranny of Oil* (2008); and *The Bush Agenda* (2006).

Fenton Lutunatabua is a Fijian storyteller, writer, photographer, organizer, and facilitator. He is the host of the podcast *Beyond the Narrative*. Currently Fenton is the head of regions at 350.org, serving Africa, Asia, Europe, Latin America, North America, and the Pacific. He has worked with the Pacific Climate Warriors since 2014 and currently serves as a member of the Secretariat.

Thelma Young Lutunatabua is a digital storyteller and activist. She is the cofounder of Not Too Late and currently works at The Solutions Project. Thelma has worked in various roles around the world in support of the global climate movement as well as other human rights endeavors. She calls Fiji and Texas home.

Yotam Marom is a facilitator, organizer, and writer based in New York City. He has played a leadership role at Occupy Wall Street and in other movement moments, and has founded and led a number of social movement organizations. He spends most of his time as a facilitator for movement organizations and as a dad. His writing and other work can be found at www.yotammarom.net.

Denali Sai Nalamalapu is a queer, South Indian American writer, artist, and climate communicator, who currently lives in Washington, DC. She is from Maine. Her family is from Andhra Pradesh, India.

Joseph Zane Sikulu is a Tongan activist and organizer working with Pacific communities on climate and LGBTI issues. He is currently the managing director for 350.org Pacific and serves the Pacific Climate Warriors as a member of the Secretariat.

David Solnit (brother of Rebecca Solnit) is an arts organizer, puppeteer, artist, and carpenter.

Rebecca Solnit is the author of more than twenty-five books, including *Orwell's Roses*; *Hope in the Dark*; *Men Explain Things to Me*; *A Paradise Built in Hell: The Extraordinary Communities That Arise in Disaster*; and *A Field Guide to Getting Lost*. A longtime climate and human rights activist, she serves on the board of the climate group Oil Change International and the advisory boards of Dayenu and Third Act.

Leah Cardamore Stokes is the Anton Vonk Associate Professor of Environmental Politics at the University of California–Santa Barbara, where she helps lead the 2035 Initiative. She is the author of *Short Circuiting Policy*, a contributor to the anthology *All We Can Save*, and cohost of the podcast *A Matter of Degrees*. She has worked to advance federal climate policy alongside numerous activists, including those at Rewiring America and Evergreen Action.

Farhana Sultana is a professor in the Department of Geography and the Environment at the Maxwell School of Citizenship and Public Affairs of Syracuse University. She is an award-winning internationally recognized interdisciplinary scholar from the Global

South. Author of several dozen publications, her latest book is *Water Politics: Governance, Justice, and the Right to Water* (2020).

Gloria Walton is an award-winning community organizer, writer, speaker, and the president and CEO of The Solutions Project. Described as one of the country's most exciting next generation political leaders and named *Inside Philanthropy*'s new president to watch in 2020, Gloria is committed to growing and supporting frontline climate justice solutions that transform our economy and world.

Index

About Haymarket Books

Haymarket Books is a radical, independent, nonprofit book publisher based in Chicago. Our mission is to publish books that contribute to struggles for social and economic justice. We strive to make our books a vibrant and organic part of social movements and the education and development of a critical, engaged, and internationalist Left.

We take inspiration and courage from our namesakes, the Haymarket Martyrs, who gave their lives fighting for a better world. Their 1886 struggle for the eight-hour day—which gave us May Day, the international workers' holiday—reminds workers around the world that ordinary people can organize and struggle for their own liberation. These struggles—against oppression, exploitation, environmental devastation, and war—continue today across the globe.

Since our founding in 2001, Haymarket has published more than nine hundred titles. Radically independent, we seek to drive a wedge into the risk-averse world of corporate book publishing. Our authors include Angela Y. Davis, Arundhati Roy, Keeanga-Yamahtta Taylor, Eve L. Ewing, Aja Monet, Mariame Kaba, Naomi Klein, Rebecca Solnit, Olúfẹ́mi O. Táíwò, Mohammed El-Kurd, José Olivarez, Noam Chomsky, Winona LaDuke, Robyn Maynard, Leanne Betasamosake Simpson, Howard Zinn, Mike Davis, Marc Lamont Hill, Dave Zirin, Astra Taylor, and Amy Goodman, among many other leading writers of our time. We are also the trade publishers of the acclaimed Historical Materialism Book Series.

Haymarket also manages a vibrant community organizing and event space in Chicago, Haymarket House, the popular Haymarket Books Live event series and podcast, and the annual Socialism Conference.

Also Available from Haymarket Books

Aftershocks of Disaster: Puerto Rico Before and After the Storm
Edited by Yarimar Bonilla and Marisol LeBrón

Atomic Days: The Untold Story of the Most Toxic Place in America
Joshua Frank

The Battle for Paradise: Puerto Rico Takes on the Disaster Capitalists
Naomi Klein

Ecosocialism: A Radical Alternative to Capitalist Catastrophe
Michael Löwy

How We Go Home: Voices from Indigenous North America
Edited by Sara Sinclair

Kivalina: A Climate Change Story
Christine Shearer

No Planet B: A Teen Vogue *Guide to the Climate Crisis*
Edited by Lucy Diavolo, foreword by Lindsay Peoples Wagner

*Too Many People?: Population, Immigration,
and the Environmental Crisis*
Ian Angus and Simon Butler, foreword by Betsy Hartmann
and Joel Kovel

The Tragedy of American Science: From the Cold War to the Forever Wars
Clifford D. Conner

About Not Too Late

#NotTooLate is a project to invite newcomers to the climate movement, as well as provide climate facts and encouragement for people who are already engaged but weary. We believe that the truths about the science, the justice-centered solutions, the growing strength of the climate movement and its achievements can help. They can assuage the sorrow and despair, and they can help people see why it's worth doing the work the climate crisis demands of us.

We know that there are still important choices to make about climate, and we don't have to wait hoping for national politicians to act. We know individual actions matter, but in order to have effective change it will take mass action from people everywhere. We know the difference between the best and worst case scenarios matters. We know that the future is being decided in the present. We know that a lot of people are overwhelmed by doom and gloom. It is not too late. At the same time, we are not here to avoid the worst news. We are here to fortify people to face it and to try to change it.

We also believe that some of the challenging emotions we feel about the planet's climate stem from commonly held frameworks about how change works, where power resides, and what possibility looks like, and we are here to think with you about those as well. Our goal is to offer good news, perspectives, voices, connections to people, as well as good paths forward for the climate and those who care about it.

Find us at https://www.nottoolateclimate.com.